W9-CZU-424

5-25-17(1)

POLLINATR
FRIENDLY
GARDENING

© 2015 Quarto Publishing Group USA Inc.
Text © 2015 Rhonda Fleming Hayes
Photography © 2015 Rhonda Fleming Hayes

First published in 2015 by Voyageur Press, an imprint of Quarto Publishing Group USA Inc., 400 First Avenue North, Suite 400, Minneapolis, MN 55401 USA.
Telephone: (612) 344-8100 Fax: (612) 344-8692

quartoknows.com
Visit our blogs at quartoknows.com

Voyageur Press titles are also available at discounts in bulk quantity for industrial or sales-promotional use. For details contact the Special Sales Manager at Quarto Publishing Group USA Inc., 400 First Avenue North, Suite 400, Minneapolis, MN 55401 USA.

10 9 8 7 6 5 4 3 2 1

ISBN: 978-0-7603-4913-7

Library of Congress Cataloging-in-Publication Data

Hayes, Rhonda Fleming, 1956- author.
  Pollinator friendly gardening : gardening for bees, butterflies, and other pollinators / Rhonda Fleming Hayes.
  pages cm
  ISBN 978-0-7603-4913-7 (sc)
  1. Gardening to attract wildlife. 2. Gardening to attract birds. 3. Bees. 4. Butterfly gardening. 5.
Hummingbirds. 6. Pollinators. I. Title. II. Title: Gardening for bees, butterflies, and other pollinators.
  QL59.H39 2016
  638'.5789--dc23
                                        2015020836

Acquiring Editor: Thom O'Hearn
Project Manager: Jordan Wiklund
Art Director: Cindy Samargia Laun
Cover Design: Karl Laun
Book Design and Layout: Amelia LeBarron

Printed in China

POLLINATOR FRIENDLY GARDENING

GARDENING FOR BEES, BUTTERFLIES, AND OTHER POLLINATORS

RHONDA FLEMING HAYES
FOREWORD BY P. ALLEN SMITH

Voyageur
Press

TABLE OF CONTENTS

FOREWORD

Colony collapse disorder, neonicotinoids, and varroa mites aren't topics one expects to chat about at a cocktail party, but these words have indeed made their way into casual conversation of late. It seems a lot of people (not just beekeepers and gardeners) are talking about pollinating insects and how their dwindling populations are impacting everything from wildlife diversity to food production.

Pollinating insects are a matter that everyone should be talking about *loudly*. This oft-overlooked workforce helps produce one-third of everything we eat—essentials like tomatoes, almonds, and apples—which makes the pollinator crisis relevant to all. And it *is* a crisis, but the good news is that you can make a difference in your own backyard.

One of the top reasons for pollinator decline is habitat loss. As humans spread out and build, we destroy wilderness where insects thrive. Common sense says that if we make our yards havens for bees, moths, birds, and other beneficials, we can return some of what we've taken away. And why wouldn't you want to invite these creatures into your garden? Not only will they bring the landscape alive, your plantings will prosper under the care of our winged friends.

I knew this book would be good when I read this sentence in the first paragraph: "It starts with changing people's mindset about 'bugs.'" This is the cornerstone in our effort to save the pollinators. The old way of thinking is that all insect pests should be completely stamped out. Not only is this impossible, but the consequence of attempting to do so is the loss of good bugs. A more positive approach is to recognize that insects are valuable components to a healthy garden's ecosystem. *All insects*. The key is to encourage the beneficials and manage rather than try to eliminate those that cause damage.

Pollinator Friendly Gardening is an excellent resource for both inspiration and practical know-how in your journey toward becoming an advocate for pollinators. It's no surprise to me that Rhonda is able to offer her wisdom in a way that is easy to understand, achievable, and without fuss. She's attended several of our annual Garden2Blog events at the farm, and I always find her passionate yet flexible, and encouraging rather than preaching. These traits are reflected in her writing: she never admonishes the reader for loving non-native plants, invites us to weave pollinator-friendly plants throughout the garden rather than relegate them to a restricted area, and assures us that we don't have to sacrifice good design when planting native species.

Because you are reading this book, you are already on your way to creating a space for both you and wildlife to enjoy. Rhonda will lead you in the right direction with plant suggestions, design ideas, and the knowledge you need to help pollinators flourish.

P. Allen Smith

Red admiral butterflies work efficiently over the tiny florets of these alliums.

INTRODUCTION

As I went through the checkout line at Whole Foods the other day, I noticed bags of "pollinator-friendly" almonds for sale. I smiled to myself. Who would have predicted that product ten years ago, or who would have known what that slogan even meant? There has been a seismic shift in attitude toward the creatures responsible for a huge portion of our food supply. Champions of organic food and agro-friendly food, farm, and garden practices are making people more aware of pollinators and the threats they face than ever before, so it's a crucial time to build on that message and encourage the creation of pollinator-friendly gardens. It starts with changing peoples' mindset about "bugs."

I've been gardening all my life, though I wasn't always cognizant of the contribution pollinators made to the process. You might say my interest and concern for pollinators came about by happy accident—although I like to think my gardening practices were already evolving in that direction, it was one random moment that put this passion into focus.

One day I sat down on the bench to rest after a sweaty session of weeding in my Kansas kitchen garden. I looked around and wondered about the garden's future now that the kids would soon leave for college. This garden had been the epicenter of my horticultural and maternal urges—the need to grow and cook—for ten years.

It was the longest I had ever gardened in one location. The picket-fenced plot was the culmination of my gardening ambitions after a lifetime of planting in fits and starts, first with a nomadic family, and then later because of my husband's job, a fast-paced position that transferred us around the country and overseas.

Our time in England proved transformative for my garden aspirations. Sure, there were the spectacular estate gardens, and everyone seemed to have beautiful perennial borders or at least fabulous hanging baskets. But it was their humble kitchen gardens that stole my heart. All that talk about bad British food vanished when newfound friends offered up gifts from their gardens: a sack of baby potatoes still crusted with bits of earth; a bundle of asparagus; a punnet of soft, sweet, no-sugar-needed strawberries. I knew when I got the chance to garden again *I wanted what they were having*.

Arriving back in the US, I went to work crafting my own version of this artful, edible landscape. The neighbors wondered aloud as the foursquare beds appeared in our front yard. Was it a pet cemetery, a formal rose garden? What, no hostas? When they saw I planned to feed my family from this very visible veggie garden, opinions were mixed. I ignored them and set about on a steep learning curve.

Determined to do well with this pretty plot, I devoured gardening books and became certified as a Master Gardener. There was so much still to learn about soil, seed starting, fertilizers, and pest control. Any mention of wildlife or insects in the garden was framed in negative terms: greedy birds, ravenous rabbits, voracious worms, stinging wasps, and those troublesome ants. There were brief discussions on pollination in connection with fruit trees, but much more detailed information on fungus, rust, scab, and galls.

Trees such as willow, ash, cottonwood, and birch are larval hosts for butterflies.

I'd never really thought much about insects or pests in the garden before. I was a keen observer of nature but I had never ventured an interest in invertebrates. When I thought of wildlife in the garden, my thinking mostly centered on larger living things—birds, turtles, the occasional chipmunk, perhaps a few conspicuous butterflies.

Back on the garden bench I was still pondering where I would direct my need to nurture with this empty nest. I paused and heard a humming sound and glanced over at a number of bees buzzing as they worked intently on the flowers I had planted between rows of kale and peppers for ornamental effect. A few butterflies fluttered, too. It was like I had never really seen them before. I hate to use the term *aha moment,* but it was indeed that.

Wanting to learn more about the role of insects in my garden, I came upon Sally Jean Cunningham's *Great Garden Companions,* a book written by another Master Gardener. Though the book was about companion planting, more strikingly it introduced me to the world of beneficial insects.

I had read other books about companion planting, but found the tales more colorful folklore than fact. With respect to generations of gardeners before me, I wanted to know the science behind the theory, when the concepts really worked, and when they were based solely upon anecdotal evidence. As I grew into my role as an Extension Master Gardener, I had grown to appreciate research-based gardening information. With so much at stake—the investment of time, perspiration, and money—I wanted my gardening efforts as well as others' in the community to always be a successful, sustainable experience.

I added more blooming plants to my kitchen garden to specifically attract *beneficials*—the tiny wasps and flies, lumbering beetles, and roving spiders would keep garden pests in check while adding a certain liveliness to my food-growing efforts. Beneficial bugs proved to be the gateway drug that fueled my passion for pollinators. I embraced the idea (contrary to conventional thought) that a healthy, productive garden shouldn't be bug-free but rather have lots and lots of these creatures performing countless helpful tasks for free.

It was a short leap from beneficial insect believer to bee watcher. Instead of seeing generic bees, I began to differentiate honeybees, fat bumblebees, iridescent green sweat bees, and many others. Butterflies, too, floated over the garden, lingering longer to drink nectar and lay eggs. I discovered this entire microcosm of a much larger habitat whirring right along inside my picket fence.

My yard faced a well-trafficked road that was a popular shortcut to the mall. People in passing cars began to yell at me, but they were nice things such as, "Love your yard," and "I pass here on purpose." Overall, the more I made the yard better for wildlife and pollinators, the more beautiful and interesting it became.

And then we had to move again.

A few years ago I saw a picture of our old house on Facebook, but something didn't look right. Friends hadn't had the heart to tell me about the changes, however, and I learned the new owners had built a second garage *directly on top* of my kitchen garden. They had also torn out many of the pollinator-friendly perennials and shrubs I planted. It was a personal example of habitat destruction, the kind that pollinators face every day on a much grander scale.

It seems there's always more discouragement:

- My favorite place to walk the dog—a scrubby lot full of wildflowers where we flushed countless fluttering butterflies—has fallen to development. What's left beyond the building's footprint is now covered in asphalt and narrow strips of sod.
- Concerned calls come into to the Extension hotline, and homeowners ask *Where are the bees?* The buds on their apple trees languish without pollination.
- There are reports of bee kills in neighborhood hives due to pesticide drift.
- Driving through the country, my husband remarks how clean the farm fields are compared to when he was a boy, now planted to the very edge and weed-free.

And that's just a few incidents in my world. I'm sure you can cite examples of habitat destruction and pollinator loss in your neighborhood as well.

Multiply incidents like those thousands of times over. Shortsighted development, damaging farming practices, irresponsible pesticide application, and other issues permeate our communities and gardens, and it's easy to see we are crowding out or killing keystone species crucial to the functioning of entire ecosystems. Then throw in disease and parasites that also threaten their existence. Bees and other pollinators may be small in size, but their significance and our interdependence is beyond scope.

As I write this book, it seems the world is filled with so much brutality and uncertainty. It's easy to feel overwhelmed and helpless in the face of so many big issues. Next to climate change, world hunger and war, gardening can seem like a frivolous activity.

Yet pollinator conservation is one global issue gardening can reshape in a huge way. No one garden can solve the pollinator problem, but small changes in thousands, maybe millions of gardens can have an enormous impact when you choose to support these vital creatures. Welcoming pollinators will not only add vigor to your food and flower gardens, it will fill your life with color, sound, movement, and most of all, joy.

HOW TO USE THIS BOOK

Many recent books about pollinators read like entomology textbooks or native-plants-only manifestoes. They don't always take into account gardeners' busy lives or everyday gardening realities.

This book aims to inspire and educate people who want to make meaningful changes, no matter how small or grand, and to help welcome and support pollinators in their gardens. Although it's a great guide for beginning gardeners as well as seasoned ones, it's not just for gardeners:

- Parents, teachers, youth leaders, and homeschoolers will find it a valuable resource that translates pollinator conservation into basic concepts that can be easily included in student curricula.
- Backyard naturalists, citizen scientists, and birdwatchers will find the information helpful to their observations and activities.
- Church groups can use the knowledge for community garden and environmental stewardship programs.
- Homebuilders and architects can learn more about landscaping geared to pollinator preservation.
- For those working on a larger scale, the book can be a leaping-off point for city councils, homeowners' associations, and even farmers.
- Golf course and cemetery groundskeepers may find it useful.
- Who knows—maybe beekeepers might learn a thing or two they didn't know before?

Pollinator Friendly Gardening offers three basic principles to support pollinators in your garden: provide food with blooming plants throughout the season; allow for nesting and overwintering sites; and finally, avoid pesticide use. By making the science behind pollinator behavior accessible, this book simplifies the steps to creating healthy habitats for these important creatures.

This book is intended for folks who want to help pollinators without necessarily ripping out their landscapes and starting over as some would demand. Use it to evaluate your existing landscape. Chances are you're doing many things right already. Then follow as many suggestions as you want to build upon that success and further improve pollinator conditions over time as your schedule and pocketbook permit.

It's for people who want to plant for pollinators without offending the neighbors or violating homeowner covenants and local laws. Many may worry that attracting bees to their property or using native plants will upset the neighbors who prefer manicured shrubs and lawns. *Pollinator Friendly Gardening* offers appealing design suggestions that merit praise rather than HOA fines. It gives information about bee identification and behavior that when shared can hopefully calm the neighbors' fears.

This book seeks to encourage pollinator conservation by offering easy, practical tips for enhancing your garden habitat so that it's enjoyable for both you and the vital creatures that share the space. *Pollinator Friendly Gardening* doesn't judge—there's no pressure to plant outside your comfort zone. Even a more conventional garden can still sustain pollinators with a bit of thought and ingenuity.

Pollinator Friendly Gardening presents effective alternatives to the pesticides, insecticides, fungicides, and herbicides that endanger pollinators. It also identifies the criteria for when their use is warranted and describes rules for subsequent safe application. It hopes to help gardeners untangle the issues with invasive plants. Learn when weeds are worrisome and when they can be wonderful.

The plant lists read like a catalog of greatest hits in support of bees, butterflies, and hummingbirds. Everything from the best bee perennials to the best trees for butterflies (trees! who knew?) and beyond provide hundreds of potential plants for your pollinator gardens. And that's just a starting point!

Resources are provided for people seeking more specialized information, as well as a vendor list of businesses selling bees and accessories and pollinator-friendly seeds and plants.

Last, *Pollinator Friendly Gardening* is a call to action. Hopefully you'll be inspired not only to support pollinators in your own garden but in the greater community. It provides examples of citizen science programs around the country adding to the buzz for protecting pollinators, some literally sowing the seeds of change upon their local landscapes.

Whether you plant a few extra flowers this year or join a movement, when it comes to pollinators, every effort helps.

Bees need flowers for the nectar and pollen they provide in return for pollination services.

CHAPTER 1

UNDERSTANDING POLLINATORS

WHAT IS POLLINATION?

Imagine a summer without blueberry pie, without icy, cold watermelon. Not into sweets that much, you say? Picture your tortilla chips minus the salsa and guacamole. If that doesn't worry you, this one will: how would you get along without that mid-afternoon chocolate bar or your morning coffee?

Without pollinators, these delicious treats would disappear. Oh, you'd still have corn and wheat, although that might get a bit dull. Pollinators are responsible for every third bite of food you take, but more importantly the colorful and healthy fruits and vegetables that perk up your dinner plate. But it goes deeper than that. Two-thirds of the entire world's plant species depend upon animal pollination. Plants that feed insects that feed birds and frogs that feed the snakes and owls and on up the food chain…you get the idea. Without the vital services of pollinators, the whole grand scheme falls apart.

It could be said that insects run the world. Yet many people still don't realize the critical role pollination plays in maintaining human sustenance and a healthy, diverse ecosystem. Some possess a vague memory from biology class, a diagram of flower parts and something about bees. Mention *pollen* and their first thoughts go to the invisible irritants that float in the air, stuffing up their noses and making their eyes itch and water. Pollen is that pesky yellow dust on their cars when they park outside. Pollen is the orange stain on their shirt when they brush up against lilies in a hotel lobby bouquet.

Far from a nuisance, pollen is the magic dust that makes everything possible. Gardeners marvel at tiny seeds and how they produce such a beautiful variety of plants. Pollen is fascinating, too. It comes in many shades besides yellow—pale gray, light green, brick red, steely blue, black, and many gradations in between. Beekeepers can

Much of this tasty lunch was made possible by pollinators.

often learn where their bees forage by noting the pollen color they bring back. You'd be surprised at some—white snowdrops, for example, have red pollen, and red poppies have black.

Pollination is the transfer of these pollen grains from the anther of one flower to the stigma of the same or another flower. This helps the plant to successfully reproduce. Some plants are wind-pollinated; a few are even pollinated by water, but most depend on insects, birds, and a few other animals.

Plants that use the wind to reproduce throw out huge quantities of lightweight pollen grains (the ones most responsible for all of those miserable symptoms) that fly through the breeze. Many trees, such as willow, birch, walnut, conifers, and even grasses rely on this system, one that takes advantage of an abundant natural resource without expending energy on producing conspicuous flowers to entice insects. Their flowers don't produce nectar and have little to no fragrance. Their stamens, like those on birch catkins, are exposed to make it easier to catch pollen passing by. With this scattered shotgun approach, though, enormous amounts of pollen miss their mark and are wasted in the process.

Insect pollination, on the other hand, is highly efficient and accurate. Pollen grains are held on the anthers at the center of the flower. When bees, butterflies, and other pollinators visit the flowers looking for nectar, they brush against the flower's anthers, catching the pollen grains on their bodies. Bees also seek their share of the pollen, packing it into specialized structures on their hairy bodies to transport back to their hive. As pollinators move from flower to flower, some of the pollen falls off and sticks to the stigma, the prominent female flower part that serves as the entrance to the flower's ovaries. The pollen grain is made up of two cells: one forms a pollen tube which then directs the other generative cell taking the pollen down into the depths of the flower where it fertilizes the waiting egg. Once fertilized, this enables the plant to make seeds.

Within this process there are different types of pollination. In self-pollination, pollen moves within one flower or between flowers of the same plant. These plants are described as self-fertile. Cross-pollination occurs when pollen moves from one plant of the same species to another. Self-fertile plants may indeed produce flowers and fruit but lack the genetic diversity that comes from repeated cross-pollination. This genetic diversity helps plants adapt to changing conditions, pests and disease, and other stresses, a quality that becomes more and more important with climate change.

TOP: Bees collect different shades of pollen depending upon the flower source.

BOTTOM: Pollen sticks to this long-horned bee's hairy body. It's then transferred as the bee moves from flower to flower.

THE POLLINATORS

HONEYBEES

Honeybees are native to just about everywhere other than North America—Europe, Africa, the Middle East, and Central Asia. The European colonists introduced them to the New World in the early 17th century, and since then they have made themselves at home and become a vital part of our agriculture—both for honey production and crop pollination. They vary from gold to brown to black. Most importantly, though, they all are fuzzy, which is ideal for collecting loads of pollen that they store in their pollen basket, a specialized body part called a *corbicula*, as they buzz back to the hive.

As social insects, honeybees live in a colony organized in castes that determine each bee's job in the hive. The queen is the largest bee in the colony and solely responsible for laying eggs. The males or drones are there for mating purposes only. In this female-driven society, the worker bees industriously perform a number of duties in a hierarchy based on age. The youngest clean the hive, nurse the brood, and attend to the queen. The next oldest workers help to guard the hive entrance. The very oldest workers forage for nectar and pollen.

Honeybees are generalist foragers, meaning they visit a variety of flowers. The location and quality of flowers they find is communicated to other workers through pheromones and an intricate "waggle dance." They are not adapted to every plant they encounter because they are a foreign species, though, and sometimes practice nectar robbing. This involves tearing a slit in the side of a flower to extract the nectar without having to enter the bloom or pollinate the flower.

TOP: Lemon chess pie with blackberry compote, courtesy of cross-pollination.

BOTTOM: A beekeeper inspects his bees and the honey frame from their hive.

NATIVE BEES

Bumblebees (49 species in US): Most cartoon bees are based off the likeness of bumblebees, making them the most beloved and recognizable of all bee species. Yet they only account for 1.4 percent of all bee species in the US. Fuzzy and chubby, they are usually black and yellow, although some can have orange, brown, or even white bands. They are generalists and forage on a wide variety of flowers. Bumblebees are able to shake pollen from a number of flowers that honeybees can't access. Using "buzz pollination," they grab the flower and vibrate their wings at a high frequency until the pollen falls from the blossom. They are important pollinators of tomatoes, watermelons, and blueberries, among other food crops.

Carpenter Bees (36 species in US): These bees come in a range of sizes. The bigger bees, such as the common Eastern Carpenter Bee, are often confused with bumblebees; however, their shiny black abdomens distinguish their species. Their big green eyes are notable, too. People are not always happy with carpenter bees' habit of chewing into soft wood for nesting purposes.

They are generalists in their foraging, valuable pollinators in the vegetable and flower garden, up and out at work early in the mornings. Like bumblebees, they can use buzz pollination. They are known to practice nectar robbing (see page 15) when they can't fit their large bodies into some blooms.

Cuckoo Bees (499 different species in US): The jury is still out on whether these bees provide much, if any, pollination service to the flowers they visit. As parasites that lay their eggs in other bees' nests, they have no need to gather pollen. Researchers are studying whether some pollen manages to adhere to their nearly hairless bodies (possibly through static attraction), therefore contributing to pollination.

Digger Bees (332 species in US): Most information about digger bees addresses how to kill them rather than how to conserve them. People find their bee-dug burrows distressing, yet they are valuable garden pollinators.

HOW BEES MAKE HONEY

Honey making is a mystical process, a combination of chemistry and hard work where the end product is something like liquid gold. Honeybees are the only bees that manufacture enough honey to harvest because they're tasked with making enough honey for the hive to eat over winter. Bumblebees, on the other hand, make only a mere spoonful since their colonies don't overwinter. Many other bee species have no need or mechanism to make honey.

The honeybee foragers—all females, by the way—have the duty to seek out flowers for gathering nectar and pollen. On a single trip, a forager will visit from 50 to 100 flowers, making up to 30 trips a day. And yet she will only produce around a tenth of a teaspoon of honey in her entire lifetime. Amazingly, it takes *two million* flowers to make a pound of honey!

As the foraging bee visits each flower, she uses her tube-shaped tongue to suck nectar. While some of the nectar fuels her with energy for flight, the rest of the nectar goes into her crop, or "honey stomach," that acts like a nectar backpack. The nectar is kept separate in the two stomachs by something similar to a double check valve. Her crop can hold up to 70 mg, however, a common load is around 20 to 40 mg. Still, that's an enormous payload considering the average weight of a worker bee is 80 mg.

Back at the hive, she transfers the nectar mouth-to-mouth to waiting worker bees. They "chew" on the nectar, adding enzymes that break down the complex sugars into more digestible simple sugars. These enzymes make it less likely to spoil. Then they spread this liquid over the combs. The nectar begins with 60 to 80 percent moisture content, but as water evaporates from the nectar, it thickens into syrup. The bees fan it to dry it out even more, down to an average of 15 percent moisture content. Meanwhile, they add other enzymes that protect it from mold and bacteria. Once it has reached the right consistency, they seal the cells with wax so it doesn't ferment. This is what gives honey such an extended shelf life. The honey is stored this way until they are ready to eat it throughout the winter. A hive needs 120 to 200 pounds of honey to make it through winter.

Sweat Bees (287 species in US): People are familiar with these bees for an unfortunate reason. As the name implies, these bees are attracted to human sweat for its salt content. In spite of their annoying behavior, they are actually important pollinators of sunflowers and watermelons, in addition to many members of the daisy family. Females carry pollen in structures called *scopae* on their hind legs.

Green Sweat Bees (29 species in US): Contrary to the name, these bees are not attracted to human sweat. They are small but striking in color, either vivid metallic green or blue. They are foraging generalists, but their short tongue limits them to smaller flowers.

Mason Bees (197 species in US): Mason bees are abundant and widespread throughout much of the US. The blue orchard bee is one of the most popular members of the genus, prized for its beautiful blue appearance and even more so for its value as an efficient pollinator in orchards. They are commercially managed in agriculture as an important pollinator of fruit and nut trees, such as apple, cherry, plum, peach, pear, and almond.

Mason bees can be purchased, and you can enlist their gentle help in home gardens and orchards as well (see Resources, page 170).

Leafcutter Bees (136 species in US): Their upturned abdomens give leafcutter bees a comical look. You'll sometimes notice their presence by the circular cutouts they leave in foliage and flowers around the garden. (Don't worry—their damage is only cosmetic.) They use the leaf and petal pieces to line the brood cells of their nests. Leafcutter bees are generalists, but some species specialize on members of the pea and aster families.

Carder Bees (64 species in US): Wool carder bees, as they are known, comb or card the downy fibers from fuzzy-leaved plants. If you see threadbare patches on your lambs' ears, it's probably evidence of a carder bee gathering wool for her nest. They are often mistaken for wasps due to the yellow and black broken banding on their abdomen. The males can be territorial in guarding the nests.

Mining Bees (466 species in US): People usually encounter mining bees in their lawns where they nest underground, but homeowners have little to fear since mining bees' stingers are weak. These bees are black,

TOP: A bumblebee uses sonification, or "buzz pollination," to release pollen from a tomato blossom.

MIDDLE: It's a tight squeeze for big carpenter bees in these penstemon flowers.

BOTTOM: Green sweat bees are striking insects with their bold stripes and emerald coloring.

dull blue, or green with a velvety patch between their eyes and around their antennae. They are some of the earliest to emerge in spring. They are both generalists and specialists in their foraging habits.

Cellophane Bees (99 species in US): Cellophane bees are named for the cellophane-like material they secrete to line and waterproof their nests. For the same reason, they are often called polyester bees. Sometimes mistaken for wasps, they are not as hairy as other bees. A number of the species transport pollen in their crops instead of outside their body.

OTHER POLLINATING INSECTS

Butterflies

Butterflies are measurably less efficient than bees at the business of pollination. Yet they may be the most popular pollinators because of their gorgeous colors and captivating flight. Their tall, slender bodies aren't conducive to picking up much pollen. With their long proboscis, however, they can reach into flowers many bees can't access. And unlike bees they are able to see the color red. For this reason they are often seen at brightly colored red, purple, and pink flowers that bees may pass by. They are important pollinators of wild and cultivated flowers, especially aster, goldenrod, dogbane, zinnias, and dahlias.

Moths

Moths work the night shift, searching out flowers that open in the late afternoon to evening especially for them. Many of these drab creatures are attracted to similarly dull-colored blooms in white, pale purple, and pinks that reflect moonlight. They are lured in further by the heavy fragrance that promises copious amounts of nectar. Of the moths that show their faces in daytime, large hovering hawk moths are well known as doppelgangers for hummingbirds. Unfortunately, their larvae—the tomato hornworm—are infamous for the damage and destruction they wreak on tomato crops. On a better note, many of these winged, nocturnal creatures are responsible for pollinating appealing plants, such as night-blooming jasmine, gardenia, yucca, and brugmansia.

TOP: The buckeye butterfly sports distinctive markings.

BOTTOM: Flower flies look very similar to bees on purpose. They use mimicry for protection from predators.

Flies

Despite their negative connotation, many members of the order Diptera are economically important pollinators of annual and bulbous ornamental flowers. Tachinid and syrphid flies, also known as flower flies or hoverflies, mimic bees and wasps in look and sound for self-protection. In fact, one looks so much like a honeybee, it's called a drone fly!

Look closely and you'll see that hoverflies act like tiny helicopters, visiting flowers with a dart-and-hover technique quite different from the more methodical bees. You'll see them nectaring shoulder to shoulder with other various bee species as well. Beyond their worth as a pollinator, their larvae are revered as voracious predators of troublesome pests such as aphids, scale, and thrips. Other species of flies are attracted to odors of rotting meat, blood, sap, and dung. North American native plants, such as pawpaw, skunk cabbage, and jack-in-the-pulpit, emit this putrid smell to attract pollinating flies.

Wasps

With such shiny bodies, wasps aren't usually thought of as good pollinators. While nowhere near as efficient as most bees, they do visit flowers and leave with a fine coating of pollen dust. The fig wasp is notable for its self-sacrificing pollination service. The tiny wasp squeezes into the immature fig fruit that keeps its flower structures hidden inside. The wasp visits the internal flowers and lays her eggs in the future seeds, spreading pollen from the fig from which it originally emerged. She never sees the light of day again.

Paper wasps congregate on this blooming *Foeniculum vulgare*, fennel.

Beetles

Beetles, by their sheer numbers, are actually the largest set of pollinating animals in the world. There are 30,000 species just in North America! They have primitive origins in the pollinating business and were the first to visit the flowers of ancient angiosperms, such as the magnolia.

In their seemingly clumsy manner, they lumber and crawl on flowers, munching on flower petals and parts, defecating as they go, earning them the title of "mess and spoil" pollinators. With this less-than-endearing behavior, no wonder they aren't actively encouraged in the garden—although they probably should be.

Beetles are drawn to flowers with spicy, sweet, or fermented smells. They have co-evolved with large, bowl-shaped flowers, such as water lilies, that offer another reward: heat. Yes, incredible as it seems, water lilies can generate their own heat. This, along with pollen, keeps the beetles coming back. Beetles and flies often spend the night folded within the petals of these cozy blooms. Of the many beetles that pollinate 52 species of native plants in North America, you may recognize the names of soldier beetles, scarab beetles, checkered beetles, jewel beetles, and tumbling flower beetles, among others.

Note: In spite of ants' notorious love of nectar, they aren't effective pollinators. They crawl up and around plants seeking out the sweet liquid; however, they also secrete a chemical that acts like a natural antibiotic that can deactivate or kill pollen grains. The flowers that they are attracted to and manage to successfully pollinate are low-growing and inconspicuous.

DR. MARLA SPIVAK

Marla's interest in bees began when she worked for a commercial beekeeper from New Mexico in 1975. She later completed her B.A. in Biology from Humboldt State University in northern California, and her PhD from the University of Kansas, under Dr. Orley "Chip" Taylor, in 1989. She spent two years in Costa Rica conducting her thesis research on the identification and ecology of Africanized and European honeybees. Influenced by Martha Gilliam and Steve Taber from the USDA bee lab in Tucson, she became interested in hygienic behavior of honeybees. This interest has expanded into studies of "social immunity," including the benefits of *propolis* (plant resins some honeybees collect and deposit back into the hive) to the immune system of honeybees, and to the health and diversity of all bee pollinators.

Q. How did you come to study bees?

I was bored and directionless in college. One day I went to the library to find a book to read. I happened to pick up a book written by an old naturalist on bees, and stayed up all night reading it. By morning I knew I had to see the inside of a beehive, and I had to see a beekeeper working with bees. So I went to work for a commercial beekeeper in New Mexico and the rest in history. I checked the book out on December 10, 1973.

Q. Are bees somewhat like humans? What can we learn from them?

I think humans try to be like bees, but bees are not like humans. Bee societies evolved millions of years before humans evolved. Honeybees are much more social than humans: an individual bee cannot live without its colony. A human can live alone his/her whole life, although we tend not to. Humans can learn something about collective decision-making and about efficient division of labor from bees, for sure. We must make sure to not destroy the habitat that other organisms (plants, animals, and bees) need for survival.

Q. What's the challenge of studying bees?

Honeybees are super-organisms: the colony is the organism and it is comprised of 30,000 to 50,000 individual bee organisms. How does the physiology or behavior of an individual change when it is alone for a limited time, versus when it is within the colony? How is the genetics or the immune system of an individual bee modified or modulated as the colony grows, or at different times of the year? Bee researchers must always be thinking at both the individual and colony level.

Q. What is the most fascinating thing you've learned about bees?

There is always something new to learn. Right now, I'm fascinated with bees' health care system—their "social immunity."

Q. Why should people care about bees?

Bees connect many of the pressing issues and offer tangible steps for remediation. Bees connect directly to our food supply through pollination of fruits, vegetables, nuts, and seeds. Providing food for bees means diversifying our landscape to ensure there are sufficient flowers from which bees can collect nectar and pollen for their good nutrition. Diversifying the landscape means changing the way we farm and maintain our agricultural and urban areas as well as paying attention to pesticide use so we don't contaminate the floral food bees depend upon. Beekeeping is a skill and a livelihood that can be done all over the world. It can provide income for small-scale farmers in underprivileged and developing countries. Products from the hive provide medicines for humans.

Q. What's the biggest threat to bees?

The biggest threat is humans. We have disturbed the landscape to the point where bees cannot live in it.

Q. Do you feel gardeners can actually make a meaningful impact?

Yes. They can plant flowers that bees like and keep those flowers free of pesticides.

Q. What's in your yard?

The front yard is native prairie. The backyard contains perennials, a vegetable garden, flowering shrubs, three honeybee colonies, and native bees that nest in the ground next to the compost pile.

NON-INSECT POLLINATORS

Birds

Hummingbirds are creatures of the New World; they only exist in the Americas. They perform deft and speedy stunts as they hover and dive around the garden, feeding on nectar as well as consuming insects. With their long proboscis, they seek nectar from deep within the tubular blooms of many shrubs and vines for which they are specifically adapted. Pollen attaches to the feathers on their head and back as they feed, and then it brushes or falls off these winged Valkyries at the next flower. The only other pollinating bird in North America is the honeycreeper, found only in Hawaii.

Bats

The next time you're drinking a margarita, pause for a moment to thank the bats: they pollinate the agave plants that are the source of tequila, the liquor with such a kick. Bats are important pollinators in tropical and desert ecosystems, working through the night to pollinate mangoes, avocados, dates, figs, bananas, and other fruits. Due to limited eyesight, bats prefer large, bowl-shaped white, pale green, or purple flowers (often detected by their musty odor). The bats' approach is awkward as they hover over the plant, then flap and lap the copious dilute nectar with their very long tongues. The pollen clings to their sticky faces and finds its way to the next flower. Although bats don't pollinate food crops in the US, there are two species of bats in the Southwest that pollinate agave, saguaro, organ pipe, and other night-blooming cacti.

Unusual Pollinators

There are a number of unusual pollinators in the world as well. Lemurs in Madagascar, honey possums in Australia, and many lizards and skinks are capable of pollination services. As strange as it may seem, gnats, slugs, mosquitos, and other bugs and insects also contribute to pollination. And you can't leave out the humans who painstakingly hand-pollinate—for example, the orchids in Africa that provide the world with coveted vanilla beans are hand-pollinated.

HOW POLLINATORS FORAGE

Find a chair and sit really still in the garden. Watch while bees and butterflies buzz and flutter from flower to flower. You might need to duck though as a hummingbird dive-bombs another fighting over a lobelia bloom! The activity all seems pretty random and haphazard, yet it's anything but for pollinators. You might even say they have it down to a science.

Study them closely and you'll see that each pollinator prefers a particular type of flower from which to nectar. Flower characteristics that appeal to certain pollinators are described as *pollination syndromes*. Put together a number of flower traits from among color, form, and scent, and it will predict the type of pollinator that will help the flower successfully reproduce.

Plants and pollinators have co-evolved for millions of years, creating a symbiotic relationship, one that benefits both parties. Pollinators find the food they need to live and reproduce while plants make sure their progeny live to set seed another day. Everyone wins.

To do this, plants have developed floral strategies that entice pollinators to visit their flowers to guarantee that pollen grains are carried to other flowers. Because flowers can't move around in their environment, they have had to develop visual and scent cues that lure pollinators to them. Large, showy flowers are an obvious come-on, as are bright colors that can be seen from a distance. Some flowers display nectar guides, dots, patterns, and stripes that function like a road map, some almost resembling

Daisy-shape flowers provide a landing pad from which honeybees can easily forage.

a big arrow pointing to the nectar prize. At other times, flowers display ultraviolet color, which is visible to bees to show the nectar location near the petals. Sometimes the methods are downright sneaky, with flowers that mimic the smell of food to bait the hook. Some devious plants ensnare pollinators with food in the form of other insects, and devour or drown them and then dissolve their bodies. (Don't ever let it be said that plants are boring.)

Meanwhile, pollinators have evolved with specialized structures and behaviors to assist in plant pollination. Bees have evolved with hairy legs and abdomens that attract and transport pollen. Bumblebees use sonication or buzz pollination for releasing difficult-to-reach pollen, such as that found in tomato blossoms.

Depending upon their foraging habits, pollinators are categorized as either *generalists* or *specialists.* Many pollinators are generalists when it comes to finding nectar and pollen. Most solitary bees are *polylectic,* meaning they gather pollen from a wide variety of flowers. This varied diet makes them more adaptable to change within their environment. Some are more selective, such as the squash bee, which only forages from members of the squash family. There are a few species of bees that are *monolectic,* though, meaning they only gather pollen from a single species of flower. Monolectic bees are most vulnerable to environmental stress with such a restricted range of food sources.

Note: Bees are the only pollinators that deliberately gather pollen. They mix it with nectar to make beebread for their brood.

PROBOSCISES FOR FLOWER PREFERENCES

Physical characteristics can determine some type of specialization. One physical adaptation that influences a pollinator's flower preference is its proboscis or tongue. Among bees, tongue lengths vary in accordance with the flower nectar they can access. Bees have a beak; two jaws or mandibles that sheath a modified tube-like tongue used for sucking up nectar. (Bumblebees have an especially hairy tongue they use like a mop to acquire nectar.) Bees fold their tongues under their bodies when not in use.

Bees have a keen sense of smell. By detecting odors with their antennae, they are able to locate flowers and other bees. They prefer mild, sweet fragrances in flowers. They are most attracted to blue, yellow, and bright white flowers, as well as those that exhibit UV color. Short-tongued bees prefer blooms that are shallow and form a landing pad from which they can feed. Long-tongued bees, however, can maneuver into tubular blossoms while balancing on flower lobes. Some bees don't want to expend the energy to gain access to complicated flower structures (those of snapdragons, for instance), so they practice nectar-robbing (described above). Other bees that come behind them often take advantage of this easy access the resulting slit offers as well.

Floral constancy is the term used when bees repeatedly visit one particular type of flower even though there might be more rewarding sources of nectar nearby. This behavior is puzzling to many experts. Are the bees just lazy, wanting a sure thing? It may be a matter of conserving energy. As far as the flower species are concerned, they benefit from this "floral fidelity" because their pollen isn't wasted on other flowers that they're unable to fertilize anyway.

WHERE HAVE ALL THE FLOWERS GONE?

Flowers have vanished from many larger landscapes, mowed down by development and modern farming practices, but flowers have also disappeared to a great extent from our front yards. Seeking low-maintenance landscapes at home and around businesses, people have turned to the lowest common denominator of greenery and installed what is called "plant material," represented on drawing boards and diagrams by static shapes and placeholders rather than showcasing their possible features and attributes. Many landscapers driven by the profitable business of building hardscape often see plants as an afterthought. Stick in a few river birches and some spiky grasses to "soften" the corners and they're good. Heaven forbid they plant something that might bloom and need tending beyond the scope of a mow-and-blow maintenance company.

Modern tastes have also played a part in the floral deficit. Foliage-driven designs figure big in many contemporary landscape schemes. It's seen as a more sophisticated look for busy people with better things to do than putter in the garden. This fabulous foliage display's continuous, unchanging color and texture without those fussy flowers is beautiful, but…. Horticulture folks often sniff at floriferous displays, denouncing them as "grandma-ish." And too often a densely planted yard labels the homeowner as eccentric.

You needn't completely forgo foliage, but remember that growing flowers is anything but frivolous, old-fashioned, or freaky. Hopefully the realization that pollinators need food and lots of it can make flower power hip again. The optimum habitat for pollinators (and in turn for other wildlife) is a season-long buffet of overlapping blooms—the more the merrier.

Butterflies, unlike bees and even unlike their own larvae, have no chewing mouthparts. Upon emerging from their pupa, their proboscis is divided into two parts that grow together in zipper-like fashion. This forms a straw with channels that draws up nectar. The long, coiled proboscis is operated by hydrostatic pressure using blood flow to coil and uncoil the long appendage so that the butterfly can insert it into flowers or even pierce fruit. Butterflies have bad eyesight, so they are dependent upon the sensors in their feet to taste and smell for evaluating flower quality. They also favor brightly colored flowers, especially reds and purples, and prefer flowers with flat landing pads that make it easy to nectar. They can also use tubular flowers that have a spur, where they can perch while feeding.

Hummingbirds have their own unique adaptation for siphoning nectar from the flower depths. Their tongues are small—less than 1 mm thick—and forked and lined with a fringe of hair-like, fleshy extensions called *lamellae*. Rather than sucking the nectar, hummingbirds lap it up in repeated quick motions with their tongue. The tips of the tongue separate as it enters the flower and the lamellae extend. Then the tips come back together as the lamellae traps the nectar and draws the liquid back in. From there, capillary action moves it down into the hummingbird's throat.

Like many other birds, hummingbirds have no sense of smell so they rely on visual cues for finding food. Many of the flowers they prefer have little to no fragrance. It's common knowledge they are attracted to bright red and orange flowers, but they also find certain white flowers appealing. They can also be seen feeding on blue flowers with the funnel shape form to which they are adapted. Even though they can hover while feeding, they appreciate a flower that provides a sturdy perch to rest upon.

HOW POLLINATORS ARE THREATENED

It's hard to know where to begin the answer to this question. A perfect storm of anthropogenic pressures looms, meaning manmade problems threaten pollinators and the vital services they provide. Most of the issues are overlapping and interconnected, not to mention overwhelming, making it difficult to list them in order of importance. But if you had to pick the top three, habitat loss, pesticide use, and climate change loom largest.

TOP: A monarch butterfly unfurls its proboscis to plumb the depths of this *Tithonia*, or Mexican sunflower.

BOTTOM: Hummingbirds are adapted to reach nectar in tubular flowers, such as *Digitalis* foxglove. Colorful dots guide the way.

HABITAT LOSS

Habitat loss comes in multiple forms with destruction, degradation, and fragmentation of the land. Intensive agricultural practices, unchecked development, and resource extraction are responsible for the loss of millions of acres of valuable habitat within the US. Farming accounts for the single largest use of land in the country, so when natural areas within farmland, farm fields, buffers, and roadsides that historically supplied food in the form of wildflowers, clovers, and weeds are mowed, sprayed, left fallow, or put into production, foraging pollinators have nowhere else to go.

Someone once described suburbs as places where they rip out the trees and name streets after them. Names such as Sunny Meadows and Pheasant Run often paint a picture of the habitat that subdivisions removed and replaced. Home landscapes with vast expanses of lawn and a scattering of shrubs and trees can't begin to restore the diverse plant communities that are wiped out in the process. You may never have seen a subdivision called Bumblebee Heights, but suburban sprawl, with its accompanying acres of concrete and other types of impervious land cover, destroys or eliminates the potential for underground burrows that native bees use to raise their young.

Open land can suffer degradation that makes it useless or even dangerous for pollinators. Soil within and around development is often scraped and compacted by bulldozers, destroying existing bee nests as well as making it unsuitable for potential nesting. Overgrazing on pasturelands not only removes needed vegetation but also puts nests and pollinators at risk of trampling. Off-road vehicles can have the same effect in parklands.

Habitat broken into disconnected patches by road construction, farming, or development isolates pollinators and other wildlife by disrupting the natural pathways they use to move through their environment while completing their life cycles. Food may be available in one area but only suitable nesting sites in another, and water somewhere else. As a result, landscape fragments caused by human activity lack diversity and tend to be "simple," with limited diet and shelter for pollinators, compared to more complex naturally occurring patches.

Habitat loss is an even bigger problem for migratory species such as the monarch butterfly and hummingbirds. They not only need abundant amounts of nectar during the summer to bulk up fat and protein supplies for their long journeys, they need a continuous supply of food along the route. When forage is sparse or interrupted by habitat fragmentation, the miraculous odyssey becomes all the more difficult.

PESTICIDE USE

Overdependence upon pesticides has had a devastating effect on pollinators. Farmers find themselves on the "pesticide treadmill" when insects quickly become resistant to one particular pesticide, necessitating more and more powerful chemicals that create superbugs and superweeds. The introduction of systemic pesticides, such as neonicotinoids that are taken up through the vascular system of plants contaminating not only foliage but nectar and pollen, are now of special concern. This insecticide may not directly kill bees, but it is detrimental to bees' ability to navigate and forage. Many of these insecticides persist in the environment with lingering negative consequences.

In the home garden, longstanding attitudes toward the presence of insects have not helped. Until recently, any information about insects always included multiple methods of killing them, even though most are engaged in beneficial or at least benign behavior. Misuse of insecticides happens regularly when they are employed without correctly identifying the target pest, with pollinators and other insects perishing as collateral damage. (They are often used at the wrong rate or time of day leading to more needless deaths.) Pesticide drift—the term for what happens when chemicals are carried astray on the wind in both field and garden—is responsible for further contamination. Herbicides may be seen innocent in the demise of insects, but in fact they can kill the food and shelter insects need to live and reproduce.

CLIMATE CHANGE

Climate change is the wild card when it comes to pollinators. Scientists can only speculate about the short- and long-term effects a warming planet will have upon pollinators and their ability to survive and continue to render their crucial services. Among many possibilities, some species may actually expand their range while others will struggle to maintain the status quo. At the same time, intricate plant/insect or plant/bird relationships that have co-evolved over thousands of years may become uncoupled. For instance, flowers and pollinators may fall out of sync, with certain flowers blooming at the wrong time within a particular pollinator's life cycle, threatening the survival of both species.

OTHER WORRIES

Invasive plants are more than just thuggish species that roam and ramble out of bounds in your yard. In the wild, invasives are usually alien plants with a penchant for world domination, displacing and crowding out the native species that provide higher-quality food for pollinators: foliage for caterpillars as well as nectar and pollen for adult pollinators. Introduced insect species brought in as biological controls for problem pests can turn into problems themselves, taking out native species. It's even been posed that honeybees compete for habitat to the detriment of native bees. It's complicated.

Parasitic mites are a major problem for managed honeybee populations. Varroa mites suck blood from adult bees as well as brood. This can result in weakened adult bees and deformed broods that contribute to an overall decline of the hive if they are allowed to become chronic infestations. Tracheal mites are another danger to bees. They clog the breathing tubes of adult bees, affecting their flight and foraging abilities, and eventually killing off large numbers of bees within the hive.

Ophryocystis elektroscirrha, or OE, is a protozoan parasite infecting monarch butterfly populations. Usually occurring during the pupal stage, OE can cause severe problems such as deformity followed by death. Even in milder cases, issues such as weight loss and weakening can be exacerbated if adequate sources of nectar aren't available to offset it.

Smog and other types of air pollution aren't seen as such a big problem in this country anymore since legislation instituting stricter air quality standards cleaned up the act of so many automobiles and industrial polluters. Current levels of air pollution, however, have been cited as a possible factor for bee problems. It has been found that scent-bearing molecules—the ones that mark scent trails for pollinating insects—are destroyed when they come into contact with ozone and other pollutants, making it difficult for foraging pollinators to find food.

Light pollution is usually considered only a problem for astronomers and their telescopes, but artificial lighting confuses pollinators who are active at night. Light, such as that from street lamps, gathers and exposes night flyers like moths to greater predation by bats, and the bats in turn are picked off by owls and other large predatory birds.

COLONY COLLAPSE DISORDER

While many fascinating mysteries surround the subject of pollinators, the story of Colony Collapse Disorder is a sad one. CCD is a baffling phenomenon where seemingly healthy honeybees abruptly disappear from their hives. The syndrome was first noticed around 2006, when beekeepers recorded unusually high losses beyond the usual winterkills. This is not without precedent—disappearances happened in the 1880s, 1920s, and as recently as the 1960s. Many speculated those earlier events were weather related; however, no clear answer exists for what is happening now.

There may be a number of issues that account for this latest vanishing act. New pathogens, such as the *Nosema* fungi, are plaguing honeybee populations. New parasites, such as the varroa mite, have weakened many colonies and forced beekeepers to resort to chemical treatments. Infestations of small hive beetles haven't helped. In addition to pressures from these pests, there is a lack of clean, nutritious food available. And it's possible to say that honeybees are overworked and exhausted; they are used extensively for pollinating an expanding amount of popular food crops, such as almonds, and transported around like cattle from orchard to orchard. Put those new systemic pesticides into the mix and it could be one or many of these stressors combined.

CAN GARDENERS MAKE A DIFFERENCE?

With such a depressing laundry list of threats to pollinators, it's easy to throw your hands up in hopeless despair. How do you support pollinators and still provide food, housing, roads, fuel, raw materials, safety, gathering spaces, and other necessities for humans at the same time?

Yet when asked whether gardeners can make a difference, experts in the scientific field answer time and time again with a resounding, "Absolutely!" In fact, gardeners may be the ones who save the day. Collectively, gardeners are in charge of a lot of land, after all.

It's understandable as well as commendable when generous and concerned people donate to causes like saving whales, rescuing tigers, and other magnificent creatures; these large, charismatic animals appeal to a broad population. Many of these animals are keystone species, meaning a species whose presence plays an irreplaceable role in maintaining the ecosystem in which it lives.

Alligators and grizzly bears are keystone species, but so are bees and hummingbirds. It's harder to rally sympathy around insect or plant conservation, yet this is a huge opportunity to save species that are the linchpin to the whole system. You can make it happen in your own backyard.

Create a pollinator-friendly garden and encourage others to do the same. They can then encourage others as well. "Plant it forward" until a series of vital habitat corridors emerge. Furnish pollinators with three things: abundant food, places to nest, and a safe environment. Then stand back and see the impressive results for yourself, in real time, right under your nose.

Native blazing star, *Liatris ligulistylis*, makes for the ultimate monarch magnet.

CHAPTER 2

INVITING POLLINATORS WITH PLANTS

NATIVES VERSUS ALIENS

It sounds like an extraterrestrial sporting event. Actually, it is a competition of sorts. Alien plants can displace and overrun native plants with negative consequences.

Alien seems like a harsh term. Sometimes these plants are referred to as "exotic," which sounds a little racy, conjuring up steamy images of hothouse orchids. Alien plants, however, are more than just houseplants gone wild. In more polite company, aliens are called "introduced" plants. Maybe it's easier to ask, what is a native plant?

Every plant is native to somewhere. For the purposes of this book, a plant is considered native if it grew historically in North America before Europeans arrived to its shores. Yet America is such a huge country with so many types of topography and climate there are even more distinctions as to what's native. A single plant species can be prolific and sweep across vast prairies or so rare it occupies only a few acres of a particular boggy swamp, yet both are called native.

Try to grow what's truly native to your garden, though, and you may find yourself severely limited to a handful of plants. Some plants brought over by settlers seem like they've been here for so long you'd think they were native. These plants have adapted and naturalized to become part of the common landscape, but that's still a far cry from native status.

Often plants travel by chance, their fate cast on the wind—literally. Birds eat seeds and drop them further afield. Seeds stick to animal hair and find their way to faraway places. They journey in peoples' pant cuffs. They mingle in burlap bags with crop seeds. Other times they arrive at their destination by design. Native Americans might have moved plants from one location to another.

Native coneflowers bloomed on the prairie long before settlers arrived in North America.

FUN FACT

Insects have been
pollinating and
co-evolving
with plants for around
96 million years.

European settlers carried seeds and cuttings with them so they could grow familiar food crops. They stowed away herbs to flavor dishes with Old World taste. Sometimes it was just for sentimental reasons, a way to transport a little bit of home in a tiny portable package.

It wasn't a one-way street. In the Golden Age of Exploration, intrepid plant hunters collected plants from the New World, risking their lives for unique specimens they could stash in trunks and barrels then load on ships bound for the wealthy and the curious in Europe and other places. Exquisite flowers and alluring leaves from the savage world seduced prim Victorians who housed and studied their acquisitions in herbariums back home. But it was more than pretty flowers; edible plants came back, too. Those quintessentially Irish potatoes and Italian tomatoes, after all, started out in the Andes Mountains of South America.

The strong desire to see how something different grows in your garden is not a new idea and not limited to one continent. The plant on the other side of the world is always greener and better, or so it was thought. The forefathers of today's horticulture business had no clue that satisfying gardeners' quest for novel and appealing plants would backfire someday.

Many of the ornamental plants imported to this country were specifically chosen for being "pest-free." The reasons they can claim this title is that insects and pathogens here haven't co-evolved with these alien plants and don't recognize the chemical compounds within their leaves as food for themselves or their offspring. Even if they wanted to, many native insects simply don't have the ability to find, eat, and digest these plants. Without a shared evolutionary history, alien plants—for all their beauty and carefree attributes—might as well be made of plastic when it comes to most hungry caterpillars and other larvae.

Both people and pollinators are drawn to camellias. Native to Asia, they were first introduced to North America in 1797.

INTRODUCED PLANTS

Wait a minute before you hang your head in shame and slink out to your garden with a shovel and shears ready to rip out all those evil aliens. Many of those alien plants are also some of the most beloved and popular plants in the garden, including all-American plants such as lilacs, peonies, hostas, roses, and hundreds of other garden-variety flowers: how could they be that bad? They aren't all that bad. Besides their beauty and classic looks they do possess habitat value. Bees, butterflies, and hummingbirds all utilize their nectar and pollen-rich blooms. And in a world where flowers are in short supply, it's better to have any kind of flowers than none.

Then there is the fact that honeybees are alien creatures themselves. So they have no problem feeding on alien flowers that hail from similar origins in Europe, Africa, Asia, and the Middle East. Introduced plants, such as herbs and old-fashioned flowers, can play an important role for pollinators as their presence in gardens helps to increase the amount of available food in general, many times filling in the gaps when nothing else is blooming.

Some plants may be recent immigrants to this country but members of a larger family of similar native plants. The easiest examples are dill and parsley that swallowtail caterpillars devour with gusto. These herbs may be of European descent, but they are cousins with the native wild carrot family. This doesn't happen often, but in this case, the chemistry is compatible enough so they are nutritionally accessible as larval host plants.

No one should feel guilty about their favorite flowers or feel they can't provide for pollinators without using native plants. Just don't rule them out, because once you get to know them you may find they are just as endearing as introduced or alien plants. Gardening in the long run is about joy and beauty, and there's no room for bullying by native plant purists. But sometimes plants can be bullies. Be warned there are times when plants go from introduced to invasive.

VIGOROUS OR INVASIVE?

No one foresaw that eventually many of the cultivated ornamentals imported for their good looks and pest-repelling capacities might escape from gardens and fend for themselves just fine in the wild. Their lack of natural enemies favors their rampant growth at the expense of native plants. Without the natural pressures of pests and disease, these plants easily outcompete native plants for available space, light, water, and nutrients. You have to go no further than one plant originally imported for its utilitarian ability to spread quickly and control erosion. Can you say kudzu? This supposedly useful plant came to overtake, dominate, and literally carpet anything in its path. While not every invasive kills natural landscapes like kudzu, at last count there were around 5,000 alien species out there displacing and destroying native plants in North America.

The long list of ne'er-do-wells always contains a few unexpected members. People are often surprised, for example, to find that English ivy, known in many places of the country as an attractive and effective groundcover, is considered a scourge in the Pacific Northwest. The problem for many "vigorous" plants may lie in that no one knows how long they need to grow and proliferate before they have the potential to become invasive in areas where they seem managed. For now, it may be merely cold temperatures keeping them in check.

Many of these plants have been overplanted in subdivisions and commercial landscapes due to their low price, pleasant appearance, and tough adaptability. Look at the list and maybe you'll see something growing happily in your front yard: Japanese knotweed, multiflora rose, autumn olive, Norway maple, Bradford pear, Chinese wisteria, Japanese honeysuckle, Oriental bittersweet, Japanese barberry, and purple loosestrife to name just a sampling.

Growing in dense, impenetrable thickets, invasive shrubs crowd out more worthy native varieties while vining invasives tend to strangle native trees—girdling their trunks and shading out foliage until they are nothing more than pitiful hulks. When birds feed on the fruits, the seeds are dispersed to more places where the greedy plants reproduce again.

Invasives threaten native species in more ways than just brute physical presence. Secondary invaders in the form of alien insects and pathogens travel ashore nestled in the soil and roots of imported plants as well. The infamous Japanese beetle showed up on the New Jersey shore in a shipment of exotic iris specimens back in 1916. Everyone knows how that worked out. The most well-known fatal diseases that decimated American chestnuts and elms and now are an increasing menace to oaks all hitchhiked here on alien plants.

THE BIG BUTTERFLY BUSH DEBATE

Well-intentioned wildlife gardeners have been planting butterfly bush, *Buddleia davidii*, for years and watching the butterflies flock to it like proverbial moths to the flame. It's not unusual to find several species of butterflies all at once, (bees, too) glued to the purple (white, red, or pink) spiked flowers, working them over like there's no tomorrow. So it's shocking to learn that butterfly bush is invasive in many parts of the country and gardeners are warned not to plant it anymore. Further, it's suggested you remove any existing plants from your garden. How can it be all that bad? It is, after all, called butterfly bush.

Before you take sides in the great Battle of Buddleia, first take in the facts. Butterfly bush (not to be confused with native butterfly weed, *Asclepias tuberosa*) already had a bad rap in certain butterfly circles since it comes from China, and you just learned that that means as an alien plant it doesn't provide food for caterpillars. Well, couldn't you just compensate by planting ten more larval host plants and be forgiven? The answer's not that easy. The problem is the profuse purple flowers produce copious quantities of seeds, up to 100,000 per plant. Born on the wind, the untold numbers of seeds readily germinate with the resulting plants colonizing stream banks, forest edges, and roadsides through much of the US, to the point of becoming invasive, even in Hawaii. The seeds can remain dormant in the soil until conditions favor their emergence. Where they aren't yet declared invasive, in many locations they are deemed noxious. Its unbridled behavior has even earned it a spot on the US Forest Service website's "weed of the week."

The plant is still not considered invasive through much of the plains where drier conditions keep it in check and into the mountains and upper Midwest where extreme cold kills it back every winter. You'll still find it on many recommended butterfly plant lists as well. And you'll see it for sale in nurseries and garden centers where butterflies feed on it right then and there, oblivious to passing customers.

In response to the invasiveness of such a popular plant, plant breeders have come up with sterile cultivars of butterfly bush. These plants only produce 2 percent of the seeds that the non-sterile cultivars make, so you can plant butterfly bush as a guilty pleasure. If you can't bear ripping out non-sterile varieties already growing in your garden, you can avoid some of the issues by following a zealous program of deadheading, making sure the seed heads aren't disposed of into the watercourse. Growing some of the smaller sterile cultivars in containers as accent pieces rather than part of your in-ground landscape habitat is another possibility.

Meanwhile why not peruse this list of butterfly bush alternatives for suitable, if not gorgeous substitutes.

Pollinators take to it like candy; however, butterfly bush, or *Buddleia*, can be very invasive.

TEN ALTERNATIVES TO BUTTERFLY BUSH

These terrific native plants are deciduous shrubs or tall, sturdy perennials that display a growth habit and/or bloom structure similar to butterfly bush yet without the drawbacks. In addition to these, there may be variations of the native species that will fit into your particular garden conditions. They are all butterfly magnets.

American boneset, *Eupatorium perfoliatum:* Zones 3—8. Fluffy white flower clusters sit atop wrinkled, green, lance-shaped leaves. Blooms late summer. Grows in full sun to partial shade, 4—6 feet tall. A cousin to Joe Pye weed. Tolerates clay soil.

Blue false indigo, *Baptisia australis:* Zones 3–9. Blue lupine-like flowers bloom atop blue-green foliage. Blooms late spring to early summer. Grows in full sun to partial shade, 3 to 4 feet tall. Striking black seedhead persist after flowering. Tolerates clay soil.

Buttonbush, *Cephalanthus occidentalis:* Zones 5–9. Glossy green oval leaves with fragrant, white pincushion blooms. Blooms early summer. Grows in full sun to partial shade, 5 to 12 feet tall. Tolerates wet soil.

Culver's root, *Veronicastrum virginicum:* Zones 3–8. White to pale blue spires on whorled, lance-shaped leaves. Blooms late spring through summer. Grows in full sun, 4 to 7 feet tall. Attracts bees as well. Tolerates wet soil.

Ironweed, *Vernonia arkansana:* Zones 5–8. Fluffy, pink-purple, tufted flowers over willowy leaves. Blooms late summer. Grows in full sun, 4 to 6 feet tall. Tolerates wet soil. Fading flowers and seedheads turn a rusty shade that explains the name.

Lead plant, *Amorpha canescens:* Zones 2–9. Light purple flower spires distinguished by gold anthers. Gray-green, pinnate compound leaves. Blooms mid- through late summer. Grows in full sun, to 3 feet tall. Tough and adaptable, tolerates dry soil.

Spicebush, *Lindera benzoin:* Zones 4–9. Unusual, fragrant, yellow-green flower clusters on oblong light green leaves. Blooms early spring. Grows in full sun to partial shade, 6 to 12 feet tall. Larval host plant for the spicebush swallowtail butterfly.

Sweet pepperbush, *Clethra alnifolia:* Zones 3–9. Showy, fragrant white panicles born above glossy dark green leaves. Blooms mid-summer. Grows in full sun to partial shade, 3 to 8 feet tall depending on cultivar. Attractive dark brown seed heads in fall. Also attracts bees and hummingbirds. Tolerates clay soil.

Sweetspire, *Itea virginica:* Zones 5–9. White drooping racemes over oval green leaves. Blooms mid-summer. Grows in full sun to partial shade, 3 to 5 feet tall.

Wild senna, *Senna marilandica:* Zones 4–9. Clusters of pea-like yellow blooms over feathery compound foliage made up of oval leaflets. Blooms mid-summer. Grows in full sun, 3–6 feet tall. Interesting black seedpods for fall. Tolerates clay soil.

Culver's root, *Veronicastrum virginicum*, pictured on the right, serves as an alternative to butterfly bush.

What it comes down to is: do your research when purchasing plants. There's no reason for ignorance; it's easy to search out invasive species. Contact your local extension for the latest information if you're unsure. Before you fall in love, check to see if a certain plant's come-hither good looks come at a price. Even if it's invasive one or two states over, that's a foreboding message. There are so many plants available to gardeners; surely a beautiful alternative exists.

NATIVARS, YEA OR NAY?

Nativar is a relatively new term used to describe cultivars selected or bred from straight native species. Sometimes you'll hear them called "near natives." Nativars are selected from native plant species or hybridized between two separate plants. In the process, plant breeders grow out a large number of a particular species and then look for variations within the group. They'll look for desirable traits such as a different flower or leaf color, or larger flower size, for instance. They search for plants with longer bloom periods or faster growing times. They might seek out those with a certain growth habit, a different shape, or perhaps a more compact size. They'll also look for pest and disease resistance.

The plants with the set of preferred traits are then propagated for the mass market. The plant breeders hope to appeal to customers who otherwise might resist, with this "new and improved" version of the native species. But not everyone is sold on them. There are several objections to this process and the result. One worry is that the new cultivar will be changed both in appearance and chemistry to the point that insects and other animals no longer recognize it as food, making it not much better than an alien plant as far as its habitat value.

ABOVE: *Baptisia* 'Screamin' Yellow' is one type of nativar, selected from *Baptisia sphaerocarpa*, wild yellow indigo.

RIGHT: *Baptisia* 'Twilite Prairie Blues' is another type of nativar, a hybrid cross of two native baptisias.

The issue of habitat value isn't cut and dried when it comes to pollinators. Many factors can determine whether a plant still remains viable as a larval host for butterflies. Bees and hummingbirds, on the other hand, don't interact with foliage and can still utilize the flowers of most cultivars.

The other problem people see is when nativars are propagated for the mass market. Many nativars are vegetatively cloned using cuttings and tissue samples, a process that results in a nearly identical plant every time. For the purpose of propagating and selling the plants, the advantage is that the breeder can offer a consistent product guaranteed to give the consumer a plant with a particular set of attributes with little variability. Opponents are concerned that this limits genetic diversity by creating lots of plants that haven't been allowed to reproduce and evolve naturally. Seed-grown plants maintain a certain level of diversity, which is important to the long-term survival of the plant. A deep gene pool makes plants more adaptable to environmental stresses, such as pests or weather variability.

Colorful *Helenium* 'Mardi Gras' is a cultivar of native sneezeweed.

There are more seed-grown nativars on the market than most think, though. They carry more diversity in their genes, but still provide the uniformity that nurseries need to sell a dependable plant with reliable characteristics. There are numerous seed-grown cultivars of aquilegia, coreopsis, asclepias, echinacea, rudbeckia, heuchera, heliopsis, and gaillardia readily available.

In spite of this, many ecologists and native plant enthusiasts are concerned that while a small quantity of nativars may not be reason for concern, they believe there could be future unforeseen problems if gardeners come to rely heavily upon nativars to the exclusion of straight species.

Researchers are now working to figure out just how wildlife may be affected by the use of nativars in the home landscape. At the Mt. Cuba Center in Delaware, an institution dedicated to the advocacy of native plants, scientists have initiated studies to analyze the nutritional value of pollen grains in selected cultivars as well as the frequency of pollinator visits to these plants. Unfortunately, so far it's impossible to issue a blanket verdict upon nativars. Each cultivar has a set of unique characteristics and has to be studied on a case-by-case basis.

The same goes for your garden. Pollinators can show different taste preferences in different locations. Opinions as to pollinator preferences are often based on anecdotal evidence and shouldn't be the end-all of your plant-buying decisions. Observe plants in other local gardens, at garden centers, and in botanical gardens near your home to determine which plants in your geographic area draw in the pollinators.

When purchasing plants, you can tell the difference between the species and a cultivar by looking at the Latin. For example, native milkweed is labeled *Asclepias incarnata* spp.; however, a named cultivar with white flowers rather than pink shows the descriptive name "Ice Ballet" after the species. Sometimes the species is only referred to as "common."

The decision whether to plant nativars has good arguments on both sides of the controversy. Some say you should only plant a nativar if the sole alternative is an alien plant. Others think that a gardener whose interest is piqued by nativars will be more willing to plant it rather than a plant with no

GEORGE COOMBS

George Coombs holds a bachelor's degree in Plant Science from the University of Delaware and currently works as a Research Horticulturist at Mt. Cuba Center, where he conducts trials on plants native to the Eastern United States. Growing up on a family farm, he developed an interest in plants early in life. An internship at Longwood Gardens and the Chicago Botanic Garden, however, refined his general interest into a focus on horticulture. He currently gardens in rural New Jersey, not far from where he grew up.

Q. How does your work in the trial garden at Mt. Cuba Center differ from that of a demonstration garden somewhere else?

It's hands-on, very practical, gardener-focused research. We'll take a whole genus, including species and cultivars, and grow them for a period of three years looking at disease and pest resistance, horticultural attributes, how long they bloom, and their shape. We plant them the first year and water as needed to get them established. Then we let things run their course. We like to call it benign neglect. Sometimes it looks rough; we want people to realize this is okay. We're looking for the ones that are going to stand out, the easiest for maintenance. Everybody loves those plants.

Q. Pollinator value has recently been added to the evaluation criteria. What are researchers looking for?

They are looking at the nutrition of the pollen grains. Pollen from different species contains different amounts of protein, carbohydrates, and other dietary elements. This is very similar to the way we think of the nutritional value of our own food. In addition, they are performing surveys to see which plants are being visited frequently and by what types of pollinators.

Q. What's the difference between species and cultivars? Is one better than the other?

A straight species is one that's naturally found in the wild. A cultivar can be selected from that for different reasons, generally for horticultural attributes. You might pick a cultivar that has another flower color or longer bloom. And so a cultivar in itself doesn't mean it's not a species. Perhaps a purple coneflower, but it could just be shorter. With a hybrid it's generally two different species bred to create a novel combination of traits.

Q. Can cultivars lack pollinator value in foliage but still possess it in flowers?

Many flowering plants have pollinator value even if their foliage has been altered to provide greater disease or pest resistance because many pollinators survive solely on the flower product. Bees are an example; they really don't interact with foliage. However, foliage becomes very important for butterfly larvae that need very specific plants to feed on. So when gardening for butterflies and moths, it's important to have native plants that can support their entire lifecycle.

Q. How can home gardeners determine pollinator value in plants they find at garden centers? What should they avoid?

Sterile plants usually don't produce pollen, but may still often provide nectar. Double flowers are usually a good indication that petals have replaced some of the pollen-producing structures. From there, pick a wide variety of flower shapes, sizes, and colors and bloom times to provide a wide array of food to attract a diversity of pollinators.

Q. What's in your yard?

Five plants I grow for pollinators are *Agastache* 'Blue Fortune', *Monarda* 'Raspberry Wine', *Salvia* 'Caradonna', *Phlox paniculata* 'Robert Poore', and *Eupatorium maculatum*.

habitat value, and hopefully grow more curious about other native plants in the long run. In the end, you should ask how the plant contributes to the overall ecosystem of your garden. Don't be afraid to ask about the origins of the plant you intend to buy so you can make an informed purchase no matter where you fall on this issue.

DESIGNING WITH NATIVE PLANTS

The misconceptions about native plants usually stem from seeing ill-conceived "wild" gardens grown without regard to planning or placement. Although native plants are tough, dependable, and adaptable, they need thought and tending just like any other part of the garden. The strongest objectors accuse native landscapes of being overgrown, neglected, and even unsafe. Those who merely grumble feel they are messy and weedy, invasive or unstructured, and worse, simply too brown.

In many conventional landscapes, homeowners strive to grow plants that produce bloom after bloom from spring to fall, a flowerbed filled with never-changing blobs of color—more like a rug than a living landscape. They prefer plants that stay in place and keep their foliage to themselves. Pity the folks who want such a predictable plantscape. A garden with lots of native plants is anything but static and it's never boring! It's dynamic: full of life, color, movement, sound, and fragrance. It ebbs and flows. It's witness to thousands of small interactions between flora and fauna. It's subtle.

SELECTING NATIVE PLANTS

The choice of native plants for your landscape depends first upon your motives. Are you doing a restoration or simply trying to integrate more natives into your existing garden? Attempting a restoration, you should look for species that are very local to your specific geographic area and the ecosystem you hope to recreate. Pure landscapes rarely exist; there are few places that haven't been altered by some type of human activity anymore. At the least you try to emulate the spirit of the native plant community that was displaced. However, the majority of people want to start out by adding native plants as they go. Inserting native plants into your existing landscape is a good start toward creating a working food web for pollinators as well.

You should start by looking to your natural surroundings for cues as to what to plant. Is your area a mix of woodland and grasslands, prairie, forest, desert, or coastline? Each region has a certain botanical identity. Native plantings look their best when they show a sense of place. No matter how native that cactus may be to Arizona, it's going to look silly in a Rhode Island front yard. This is a big country and native is a local concept.

The architectural style of your home may play a part in the selection as well. Just as conventional gardens are organized with certain design principles, so are native plantings. You may use the same plant palette for a bungalow, a Tudor, or a rambler, but they may be arranged in a very different fashion.

More importantly, choose plants that fit the conditions of your yard rather than forcing them to adapt. Don't try to put a prairie into wet, boggy areas, and conversely don't attempt to grow moisture-loving plants in a dry, windswept part of your yard. Learn the characteristics of your soil. Is it clay, loam, sandy, or a combination? Study the drainage pattern of your yard, where water sheds quickly or where it tends to sit. Know your light, how many hours of full sun is available and where shadows fall. Once you have this information, you'll know better which plants will flourish in your yard.

The "right plant for the right place" mantra is repeated throughout this book. When you assemble your plants in the landscape, group them according to their cultural requirements in regards to light exposure, soil type, and moisture needs; this keeps them healthier and makes them easier to care for. This also creates harmonious plant combinations that go together visually.

As you add native plants to the landscape, use crisp lines to maintain a sense of order.

Specialist native plant nurseries in your area are the first place to search out compatible plants. Garden centers may carry a limited line, but it's still worth looking. Ask for more native plants each time you visit and eventually they may listen. Online native plant nurseries can be a great resource, but endeavor to still find natives that are most suitable for your area.

NEIGHBORLY NATIVES

No matter which plants you bring home, it won't matter if the neighbors aren't on board. No one wants to be that guy, the one whose front yard is the talk of the block for all the wrong reasons. Never fear, there are plenty of design and planting strategies that can keep you from butting heads with cultural expectations. People say they like natural, just not too natural. Neat and orderly is how they really like their natural.

Different landscape styles elicit different emotional responses. Not everyone will take yours to heart, but you can calm peoples' concerns by displaying what are called "cues to care." By following certain norms you can show your landscape plan is intentional. Seeing that the landscape is tended and weeded and not left to go wild will go a long way in keeping everyone happy.

Regardless of what the neighbors think, all landscapes benefit from structure. Of all the landscape conventions, people hold dear the notion of foundation plantings. Cover our house's bottom, so to speak. If your foundation shrubs are trimmed and tidy, you can get away with all sorts of stuff elsewhere! Shrubs in straight lines rather than fences can define boundaries. If compatible with your region, use several conifers or evergreens to anchor your home to its site and give a sense of permanence to your design, in addition to the cover they provide for wildlife.

Keep your plant palette simple to avoid a chaotic look. For folks starting out, it's recommended to go for a minimum of three varieties of blooms for each season, spring, summer, and fall, to provide a non-stop source of food for pollinators. Grow these flowering perennials and ornamental grasses in drifts, in groups of odd numbers, such as threes and fives. You can break these groups in a ratio of 2:1, for example, with one set a little ways apart for a less calculated look. People associate the sometimes wispy, smaller flowers of native plants as weedy, so try to incorporate some flowers with larger shapes and bolder color to balance the texture.

Keep a sense of scale, especially for smaller flowerbeds. Follow the usual rules; plant short in front and tall in back, or short around the edges and tall in the middle. Don't plant a forest of all tall plants in a small bed; they'll overwhelm it and actually make it look smaller, not to mention unkempt. This applies especially to boulevard and hellstrip plantings when done to reduce lawn area. It's important that they don't make people feel uneasy by hiding oncoming pedestrians or obstructing visibility. Make sure plants don't flop over the sidewalk so walkers and strollers can get through without difficulty.

People "read the edges" of the garden to determine the care given to it by the gardener. Whether straight or curved, if the edges are crisp lines, the plants behind it can be more relaxed. Mowed strips of lawn help to make natural plantings more acceptable to the untrained eye. Used as borders or paths, they let the eye rest as well as allow access for maintenance.

To maintain this visual peace at summer's end, be selective about which plants to trim back and those to leave up. Plants with handsome seedheads definitely should stay intact. Other plants that are standing sturdy and upright can be left up. Those with looser structure and that will fall over in snow or wind should be trimmed back to 12 to 15 inches high to still allow them to function as shelter for wildlife.

Architectural details in the landscape suggest a human presence. Containers, windowboxes, lamp posts, mailboxes, fences, and garden ornaments are all orderly elements that can tame the appearance of a wilder garden. Show the intent of your habitat by installing birdhouses and bird feeders, and suddenly it makes sense to passersby.

If you're not ready to go completely native, don't feel you have to deny yourself the plants you love to grow (or if you want to grow a cactus in Rhode Island). Use containers for those gaudier annuals or island fantasy plants and place them around outdoor dining and sitting areas as accents where you can enjoy their color and whimsy. Colorful pots at the front door are welcoming, another signal to neighbors that your home is cared for and well maintained.

Meanwhile, a small, simple (not preachy) sign explaining the purpose and progress of your native plantings is a great way to educate your neighbors and allay concerns. However, the best way to show your landscape is cared for is to be out there where they see you tending to and enjoying your beautiful, lively garden habitat. Hopefully they will be encouraged to do the same!

TOP: The bison sculpture within this prairie landscape provides a "cue to care" that shows the planting is intentional.

BOTTOM: Vintage bistro chairs lend a human presence to the wild garden. *Cole Burrell*

PLANT LISTS

HOW TO USE PLANT LISTS

Plants included on these lists are suitable for a broad range of conditions and are widely available at garden centers and mail order nurseries. There are hundreds of other suitable plants for pollinator-friendly gardens, too many to list in one place. Consider these suggestions a good place to start and a great leaping-off point as well. Verify your hardiness zone before purchasing plants; often there are more cultivars within one species that will fit your zone.

For plants specific to your region, especially those for coastal and desert regions, check out the pollinator-friendly planting guides at www.pollinator.org. Simply enter your zip code to find more plants for your specific ecoregion.

PERENNIALS FOR BEES		
COMMON NAME	**SCIENTIFIC NAME**	**COMMENTS** (N) NATIVE / (NN) NON-NATIVE
Anise hyssop	*Agastache foeniculum*	(N)
Aster	*Aster*	(N) and cultivars
Blanket flower	*Gaillardia*	(N) and cultivars
Blazing star	*Liatris*	(N)
California poppy	*Eschscholzia californica*	(N)
Catmint	*Nepeta racemosa*	(NN)
Culver's root	*Veronicastrum virginicum*	(N)
Cup plant	*Silphium perfoliatum*	(N)
Goldenrod	*Solidago*	(N) and cultivars
Globe thistle	*Echinops*	(NN)
Joe Pye weed	*Eupatorium*	(N) and cultivars
Lanceleaf coreopsis	*Coreopsis lanceolata*	(N)
Large beardtongue	*Penstemon grandiflorus*	(N)
Lead plant	*Amorpha canescen*	(N)
Purple coneflower	*Echinacea*	(N) and cultivars
Purple prairie clover	*Dalea purpurea*	(N)
Russian sage	*Perovskia*	(NN)
Slender mountain mint	*Pycnanthemum*	(N)
Sunflower	*Helianthus*	(N)
Swamp milkweed	*Asclepias incarnata*	(N) and cultivars
Turtlehead	*Chelone*	(N) and cultivars
Virginia waterleaf	*Hydrophyllum virginianum*	(N)
Wild bergamot	*Monarda fistulosa*	(N)
Wild geranium	*Geranium maculatum*	(N)
Yellow coneflower	*Ratibida pinnata*	(N)

Steely blue globe thistle, Echinops, is a popular perennial for bees.

ALL SORTS OF SEDUMS

Sedums are recommended for butterfly gardens, but bees are really their biggest fans. In late summer they can be seen working diligently over the tiny florets of the cushiony blooms. Although bees visit all the numerous types of sedums, they particularly enjoy the taller varieties with the large, crowned flower clusters, such as the beloved 'Autumn Joy'. It's a bonus that these great plants are very drought tolerant once established.

Sedum, 'Matrona'; Zones 3–9, Pale pink flowers on gray-green foliage

Sedum, 'Purple Emperor'; Zones 3–7, Dusky pink flowers on purple foliage

Sedum, 'Neon'; Zones 3–9, Rosy magenta flowers

Sedum, 'Frosty Morn'; Zones 3–9, Variegated foliage

Sedum, 'Xenox'; Zones 3–9, Purple foliage

Sedum, 'Rosy Glow'; Zones 3–9, Ruby red flowers

Sedum, 'Class Act'; Zones 4–9, Bright rose-pink flowers

Sedum, 'Dynomite'; Zones 4–9, Shorter, dense habit with deep rose flowers

Sedum, 'Frosted Fire'; Zones 4–8, Reddish flowers on white-edged foliage

Sedums are recommended for butterfly gardens, but bees are big fans, too.

PERENNIALS FOR BUTTERFLIES

COMMON NAME	SCIENTIFIC NAME	COMMENTS (N) NATIVE / (NN) NON-NATIVE
Allium	*Allium*	(N) and (NN) and cultivars
Astilbe	*Astilbe*	(N) and (NN) and cultivars
Bee balm	*Monarda didyma*	(N) and cultivars
Black cohosh	*Actaea racemosa*	(N)
Blazing star	*Liatris*	(N) and cultivars
Blue false indigo	*Baptisia australis*	(N) and cultivars
Bluestar	*Amsonia*	(N) and cultivars
Butterfly weed	*Asclepias*	(N) and cultivars
Catmint	*Nepeta*	(NN)
Chrysanthemum	*Dendranthema*	(NN)
Coreopsis	*Coreopsis*	(N) and cultivars
Joe Pye weed	*Eupatorium*	(N) and cultivars
Jupiter's beard	*Centranthus*	(NN)
Lupine	*Lupinus*	(N) and cultivars
Mountain mint	*Pycnanthemum*	(N)
New England aster	*Aster novae-angliae*	(N)
Phlox	*Phlox*	(N) and cultivars
Purple coneflower	*Echinacea*	(N) and cultivars
Salvia	*Salvia*	(N) and (NN)
Sedum	*Sedum*	(NN)
Sneezeweed	*Helenium*	(N) and cultivars
Turk's cap lily	*Lilium superbum*	(N)
Wild bergamot	*Monarda fistulosa*	(N)
Yarrow	*Achillea*	(N) and cultivars

ABOVE: Bees flock to the pretty purple flowers of *Geranium maculatum*, wild geranium.

RIGHT: *Monarda fistulosa*, wild bergamot, attracts both butterflies and hummingbirds.

Joe Pye weed, *Eupatorium*, lures pollinators in droves.

PERENNIALS FOR HUMMINGBIRDS		
COMMON NAME	**SCIENTIFIC NAME**	**COMMENTS** (N) NATIVE / (NN) NON-NATIVE
Bearded penstemon	*Penstemon barbartus*	(N)
Canada lily	*Lilium canadense*	(N)
Cardinal flower	*Lobelia cardinalis*	(N)
Columbine	*Aquilegia canadensis*	(N)
Coral bells	*Heuchera sanguinea*	(N) and cultivars
Foxglove	*Digitalis*	(NN)
Great blue lobelia	*Lobelia siphilitica*	(N)
Hosta	*Hosta*	(NN)
Larkspur	*Delphinium*	(N and NN)
Lupine	*Lupinus*	(N and NN)
Red hot poker	*Kniphofia*	(NN)
Royal catchfly	*Silene regia*	(N)
Scarlet bee balm	*Monarda didyma*	(N)
Virginia bluebells	*Mertensia virginica*	(N)

A carpet of Virginia bluebells, *Mertensia virginica*, and blue squill, *Scilla siberica*, welcomes the first pollinators of spring.

CHAPTER 3

ENHANCING HABITAT FOR POLLINATORS

A HOLISTIC APPROACH TO YOUR GARDEN

Too many people point to a place in their backyard and say, "There's the butterfly garden over there. . ." referring to a bed filled with beautiful blooms set aside for this specific purpose. Bless these gardeners' hearts; pollinators appreciate the effort. The problem is, the butterfly garden is just one small part of the yard and it's "over there." It's a great start: however, pollinators need more than just a patch of pretty flowers. Sure they'll pass through and sip for a second before they eventually leave to look for a better place to settle down and raise their families. Unfortunately, chances are they'll find more of the same, tastefully tidy yards with little in the way of food, shelter, and nesting sites.

This doesn't mean you have to let the yard go, leaving it "for the birds." Just the opposite: A truly successful pollinator-friendly garden embraces the total landscape, taking into account the whole yard and/ or garden from top to bottom. It's possible to provide for your practical needs while enhancing habitat, often resulting in a delightful mix of function and beauty. In subsequent chapters, you'll read how even manmade structures, walls, paths, and patios can help pollinators. Usually the solution is a simple one: more plants. Pollinators need more flowers—more food—but not just in that three-foot high zone above the ground known as the flowerbed. They need trees, shrubs, vines, and groundcovers as well, for their flowers, foliage, and other structural features. In fact, most folks are surprised to find out that certain trees and shrubs support hundreds of pollinator species, around ten times more than those perennials they planted expressly for them.

To help pollinators, you have to see your yard as they do. Across the country there are different vegetative regions; forests, woodlands, and prairies are common

This front yard includes many elements of a layered landscape with trees, shrubs, vines, groundcovers, and lots of blooming flowers.

types. The gardens surrounding homes rarely represent just one of these vegetative regions, borrowing instead elements from all of them, arranged within the confines of a conventional lot to form what is called a landscape. Back out in nature, the transitional areas where they change over from one type to another are the closest approximation to our own yards where it goes from lawn to flowerbed to shrub to tree—though in our yards it happens in short order.

It helps to think three-dimensionally about these zones or layers and how they might apply to your landscape. Starting at the top, tall trees, both deciduous and evergreen, form the canopy. Converting energy from unobstructed sun, they make huge amounts of biomass: wood, foliage, flowers, seeds, and fruit. Underneath the canopy are smaller understory trees that grow in the dappled shade and filtered light of the larger trees. Some are multi-stemmed rather than having a main trunk, and some form small groves. The next layer consists of shrubs, woody plants of various heights, often growing closely together in dense thickets. The herbaceous layer is made up of non-woody perennial plants and grasses. At the lowest level, small creeping or spreading plants hug the ground. Vines move up through all the layers ascending on rope-like stems bearing prolific foliage and flowers and perhaps fruit.

Depending upon the species, wildlife move up and down through the different layers of the landscape according to their different functions. Small mammals seek out fruits, nuts, and seeds in the treetops. Birds scout for caterpillars in the uppermost branches or pick through leaf litter on the ground for beetles, while tending their nests somewhere in the middle. The same goes for pollinators in various life stages, adults buzzing and fluttering around the flowers drinking nectar and gathering pollen while below, their larvae creep up stems and chew through foliage. Eggs and dormant larvae are tucked away in underground nests or wood cavities. All the while insects chew and bore, eat and get eaten, among thousands of other intricate plant and animal interactions taking place mostly unnoticed.

Most yards are sparsely appointed, a tree or two, some shrubs, a few flowers. It bears repeating; consider this permission to plant more plants! You're thinking that that sounds like more work on your weekend. However, adding more types of vegetation creates more biodiversity in your garden that in turn increases its biological complexity that leads to greater biological stability. Translated to plain English, this means a garden with a wider variety of plants with a working food web for wildlife is more resilient in the face of pests and diseases, more able to bounce back from weather-related issues, and more sustainable in the long run. Not to mention more beautiful.

Look for opportunities to build in more vegetative layers to your landscape. Whether it's a few more habitat-worthy bushes that fit into your budget or a grander undertaking, such as sowing a wildflower meadow, it all contributes to the cause. Creating a pollinator-friendly garden doesn't have to happen overnight. It can be a slower process of add and subtract, multiply and divide. Don't mourn for a plant when it dies; see it as a chance to add something better. If you see that one particular plant in your garden is popular with pollinators, figure out how to make more of it. Can you propagate it from seeds or simply divide it and make three more?

Or, over the next year, take out a bit of lawn and make room for smaller shrubs in front of larger shrubs. There'll be less to mow and more to appreciate. Find the perfect corner for a vine to climb. People often underestimate the number of trees their property can support. Try planting small understory trees near stately older ones and then use a flowering groundcover at their feet. Squeeze in more flowering perennials. Container gardens full of bee-baiting annuals might as well be another layer. And don't forget the vegetable garden; it can never have enough pollinators. Always be on the lookout to enrich your layers.

No matter the size of your yard, every plant can perform a vital role for pollinators, providing food, cover, or nesting sites—even the weeds! A diverse, plant-rich environment means there's a greater chance bloom times will overlap, making sure an abundance of food is available throughout the growing season. Rather than just stopping by for a visit, pollinators will take up residence, rewarding the gardener with their fascinating presence and essential services.

PERFECT HERBS FOR POLLINATORS

If you do nothing else in your yard, plant an herb garden to share with pollinators. Herbs are explained as plants cultivated for use and delight—a spot-on definition if there ever was one. Herbs are such a good investment of garden space, giving more bang for the buck and effort than just about any plant in the garden. They are easy and forgiving to grow. They're tasty and fragrant, and beautiful, too. And best of all, herbs are great for bees, butterflies, and hummingbirds.

Herbs are easy to grow but some publications will lead you to believe they thrive on neglect. It's true that established plants can survive without too much attention, but an herb garden needs careful tending at first. Pay attention to its cultural requirements: lots of sun, good air circulation, well-drained soil, and adequate but not too much water.

Plant the herb garden and let it grow for a while. Pinch the leaves for any number of uses, for cooking, floral arrangements, crafts, or even cosmetics. Pluck the foliage; crush it between your fingers, and breath in the wonderful scents. Don't be afraid to harvest them frequently. Herbs respond well to shearing and pruning, growing fuller and bushier with each cut.

Then do something for the pollinators: stop picking and snipping half of your herbs, more if you're feeling generous. Allow them to flower; it won't take long. Before you know it, bees and butterflies will flock to the blooms, and hummingbirds, too.

Pollinators love herbs in bloom because many are compound flowers, each with bunches of little florets perfectly shaped for browsing and foraging. Others bear multiples of tiny flowers in clusters or umbels. For pollinators, this means they can work over a large number of flowers within a small area, making it possible to find more nectar while conserving energy. For bees, it increases the amount of nectar and pollen they can take back to the hive with fewer foraging trips. It's all about efficiency.

Basils are always a big hit with bees.

BEST HERBS FOR BEES

NAME	SCIENTIFIC NAME
Angelica	Angelica gigas
Anise hyssop	Agastache foeniculum
Basil	Ocimum basilicum
Betony	Stachys officinalis
Borage	Borago officinalis
Catmint	Nepeta
Chamomile	Chamaemelum
Dill	Anethum graveolens
Germander	Teucrium
Lemon balm	Melissa officinalis
Mint	Mentha
Oregano	Origanum vulgare
Rosemary	Rosmarinus officinalis
Sage	Salvia officinalis
Savory	Satureja
Thyme	Thymus vulgaris

BEST HERBS FOR HUMMINGBIRDS

NAME	SCIENTIFIC NAME
Anise hyssop	Agastache foeniculum
Bee balm	Monarda
Catnip	Nepeta
Hyssop	Hyssopus officinalis
Lavender	Lavandula
Mallow	Malva
Mint	Mentha
Pineapple sage	Salvia elegans
Rosemary	Rosmarinus officinalis

BEST HERBS FOR BUTTERFLIES

NAME	SCIENTIFIC NAME
Basil	Ocimum basilicum
Catmint	Nepeta × faassenii, N. racemosa
Chives	Allium schoenoprasum
Cilantro	Coriandrum sativum
Dill	Anethum graveolens
Fennel	Foeniculum vulgare
Marjoram	Origanum majorana
Mint	Mentha
Parsley	Petroselinum crispum
Scented geranium	Pelargonium
Yarrow	Achillea millefolium

VEGETABLE (AND FRUIT) GARDENS

Without pollinators, a vegetable garden isn't possible. Sure, you can plant leaf or root vegetables such as cabbage or turnips for a harvest independent of pollinators for a season, but those seeds at the garden center have to come from somewhere. Pollinators are needed not only to fertilize flowers for fruit set, but also to produce the seeds that ensure the next generation is available to gardeners.

Even crops that are wind pollinated or self-fertile benefit from the activity of pollinators. Crops are more completely pollinated with the help of insects, a process called *entomophily*, and have increased yields, as well as bigger and better shaped fruit. Cut an apple in half crosswise. You'll see five seed pockets with a capacity for two seeds each. If there are ten seeds in the apple, it is completely pollinated. Insect pollination is essential to the development of flowering vegetables, such as squash, melons, pumpkins, and cucumbers. Tomatoes, peppers, and eggplant are wind-pollinated but also derive a boost from the work of bumblebees and other native bees also capable of "buzz pollination."

It's easy to encourage more visiting bees by adding pollinator-friendly flowers wherever possible in the vegetable garden area. You can install borders of bee-attracting perennials or intersperse annuals between rows for a consistent source of blooms throughout the growing season. Bees prefer daisy-shaped, old-fashioned flowers that can do double duty as cut flowers. Share the blooms for bouquets as well as bee bait. Rows or swaths of annuals, such as bachelor buttons, calendula, zinnias, cosmos, gaillardia, and poppies, are easy and economical to start from seed for this purpose and supply plentiful bee food from spring to fall. Shorter mounding types, such as alyssum, candytuft, French marigolds, and ageratum, can be used to outline beds and lure pollinators at the same time. Perennial borders with bee-favored cultivars chosen for a succession of seasonal blooms are a good investment when planted in close proximity to the vegetable garden.

Mix flowers with your vegetables to attract more pollinators and beneficial insects.

When warm weather arrives in the vegetable garden, many cool season vegetables, such as lettuce, radish, spinach, and broccoli, throw out flower stalks in an attempt to set seed. All plants are programmed to reproduce if they get the chance. The process is referred to as *bolting*. Resist the urge to tidy the garden and leave some of the bolted plants; they will attract a number of beneficial insects, such as bees and other pollinators. You can even save the resulting seed for next year's garden if they come from open-pollinated varieties rather than hybrids.

Once you've welcomed all these bees, be careful to schedule the watering of your vegetable garden so it doesn't interfere with their peak foraging times. Sprinklers and overhead watering simulate a driving rainstorm and create hazardous flying conditions for small creatures like bees. To avoid drenching the little guys, aim to complete watering by 8 a.m., before they venture out for the day. Don't water during night and evening hours as this can promote fungal disease.

DON'T FORGET ABOUT FRUIT

It used to be most backyards had a few fruit trees—an apple tree and perhaps a sour cherry tree for pie-making. There were berry bushes, too; raspberries or blackberries ready to fill jam jars. It's a shame that somewhere along the line people became convinced that fruit trees and shrubs are messy and troublesome, too large and sprawling, and their harvest-a-plenty too much to handle. In modern subdivisions, fruit trees and brambles don't fit into the typical one shade tree + three foundation shrubs landscape plan. In older neighborhoods, lots of leftover fruiting plants languish in an overgrown state of neglect or gnarled from disease, bad pruning, or total lack thereof.

Blooming in spring and into early summer, the bountiful blossoms of fruit trees and bushes contribute to what is called the *nectar flow*. Also referred to as *honey flow*, fruiting plants along with summer forage of clovers and wildflowers and fall blooms of aster and goldenrod make up the mainstay of food sustaining bees throughout the growing season.

Open, accessible flowers, such as those of the raspberry, rich in pollen and nectar, are irresistible to bees. Raspberries are described as self-fertile, yet without insect pollination, the fruits produce only

Fruit trees are a win-win for people and pollinators, plus they add structural complexity to the landscape.

a few nibs surrounding a corky center. Studies show time after time a vast improvement in fruit set and development when bees visit the blooms. Honeybees pollinate with precision, moving around each flower, but bumblebees and other wild bees wallow in the blooms seeking pollen more than nectar. As they move back and forth among the flowers, they actually do a better job of pollination.

Meanwhile, as honeybees are working themselves to death literally in large-scale fruit- and nut-growing operations, growing fruit at home has fallen from favor. Plant breeders and marketers have responded to some of these objections with smaller cultivars, "patio" versions of blueberries, raspberries, and strawberries that previously took up loads of space. The harvest from these new varieties may not be as prolific, but it's a start. Everyone should know the simple joy of—courtesy of a bee—a few hand-picked berries on one's Cheerios.

Hopefully, with the resurgence of interest in fresh and local food, canning, pickling, and jam-making, fruit growing may become fashionable again. Meanwhile, a glut of fruit doesn't have to mean a steamy session in the kitchen on a hot summer day. Freezing the berry harvest for breakfast smoothies or an oatmeal topping is easy; arrange the berries in a single layer on a baking sheet so they freeze quickly and without sticking together. A few hours later, throw the frozen berries in a baggy where you can grab them a handful at a time.

In other situations, maybe it's a matter of substituting a fruit-bearing plant for an ornamental with low habitat value. Berry bushes can be sited in sunny but forgotten corners of the yard growing only grass. Strawberries can work as a groundcover. Blueberries can perform like ornamental shrubs in a hedge by the driveway, displaying dense foliage and autumn color, plus bonus berries. Less common fruiting plants, such as native chokeberry, elderberry, and serviceberry, are great landscape plants that offer blooms during that crucial spring period and tasty fruit if you can beat the birds. (But if birds are the main culprits behind objections related to mess, locate those plants as far from the driveway as possible.)

Dwarf and semi-dwarf fruit trees can be used as an alternative to other small trees that have fewer landscape attributes. Smaller trees mean smaller crop yields, but busy people may not have the time to "put up" or put up with a harvest that happens all at once anyways.

And remember, for every person who doesn't know what to do with all that fruit, there is someone (without the space or conditions to grow it) who would love to have it. With social media, it's easier to find takers to share in (and even pick) a bumper crop of apples, peaches, or whatever. In California, it's not unusual to see bags of oranges and lemons sitting on the curb free for the taking. Consider donating the harvest to a local food bank; some even have volunteers who undertake the old-fashioned but newfound practice of "gleaning."

THE LAWN

No growing area gets people hotter under the collar these days than the lawn. Whether they're sweating a scorcher mowing it or debating its relative merits, things can get heated in more ways than one. Americans love their lawns—neatly mowed, of course—but lately more and more, there are signs the romance has started to sour for some. People have such strong opinions about grass that the ubiquitous sward of green may be more dueling ground than picnic spot.

Turf grass is this country's largest irrigated crop; it ranks at four times higher than corn, the next largest irrigated crop. It's pretty shocking, given that you can't eat grass. Recent studies calculate that tended lawns cover around 60,000 square miles in the US. Keeping it trim, green, and clear of weeds contributes to a $30-billion lawn care industry. That's a lot of inputs. *Inputs* is the term used to describe the resources put in to managing a crop, and in the case of lawns, that means water, fertilizer, pesticides, fossil fuels, and so forth. Not to mention all those hours behind the mower.

Some people will say that lawns are worthless in the landscape: acres of hungry, thirsty blades of grass that contribute nothing to the environment. Still, grass lawns are not inherently bad; they just aren't suited for a lot of the climate, soil, and topography where they are sowed and sodded. That leads to Herculean efforts to keep them looking like an English country estate, even in the desert.

Lawns do have their good points. Number one is that soft, cool feeling between your toes on a hot summer day. Following right behind is the scent of fresh-cut grass. More objective benefits include a reduction in noise and glare as lawns absorb sound and filter bright light. Lawns also capture dust and other air particles. They cool the air while also providing psychological cooling. They can be used as buffers for fire retardation. And in some cases they prevent water runoff and soil erosion.

Turf areas have the potential to sequester more carbon than they emit up to a certain point, hopefully countering at least some of the emissions from lawn mowing and maintenance equipment. Finally, you've got to give grass credit: it's tiny, tenacious foliage keeps returning no matter how many times you cut it.

Turf grass is the best choice for parks and recreational areas where it provides access and a safe surface for play and sport as well. In these situations, lawns provide a valuable function. The problem is that a majority of the lawns under cultivation exceed the size needed for any function. You've all seen the huge suburban homes surrounded by acres of mower-striped turf. What can planting and maintaining such a labor-intensive landscape possibly gain?

Lawns give people a sense of comfort. Possibly ingrained in our ancient DNA is "the savanna syndrome;" it explains that people felt at ease when they had a clear view of approaching predators, as well as a place to graze their animals. In modern times, especially from the 1950s on, lawns became a symbol of middle-class respectability and civic pride.

It can be said that lawns embody a sense of democracy and visually unify neighborhoods so that everyone shares in the park-like setting. Therein lies part of the problem; people get upset when that flow of perfectly shorn, velvety green is interrupted. Failure to mow is seen as blight on the community, often backed up with a notice citing a potential health menace that could harbor rats and other undesirables. When neighbors prefer a homogenized appearance to the public face of the front yard, even exuberant flowerbeds can be seen as subversive. There is a division between front/backyard acceptability, and typically, front yards have not been the most welcoming places for shows of self-expression or progressive horticulture.

Still, there is good reason now to join the growing number who ditch the turf. Besides the incredible waste of water and the scary amount of petrochemicals and pollution that follow the planting of turf grass, it really has little to offer wildlife. Very few insects or other tiny creatures actually eat the stuff. Granted,

ASK THE EXPERT:

C. COLSTON BURRELL

C. Colston Burrell is an acclaimed lecturer, garden designer, and photographer. He is the author of 12 gardening books and is a certified chlorophyll addict. Cole is an avid and lifelong plantsman, gardener, and naturalist. He is a popular lecturer internationally on topics of design, plants, and ecology, sharing his knowledge and enthusiasm with professional and amateur audiences for 40 years. He is principal of Native Landscape Design and Restoration, which specializes in blending nature and culture through artistic design. In 2008, Cole received the Award of Distinction from the Association of Professional Landscape Designers for his work promoting sustainable gardening practices. He gardens on 10 wild acres in the Blue Ridge Mountains of Virginia.

Q. What's the first plant you remember?

One is a snowdrop. My mother was a gardener, and we had those in the front yard. Then it would be a tie between hepatica and bloodroot. When I was young, we moved out to the suburbs with all these woods behind our house. It was during the time of Ladybird Johnson and beautification and wildflowers, so my mother got interested in that. We took a lot of walks in the woods. So I grew up as a gardener and nerdy junior naturalist. In addition to native plants, I became very interested in tropical plants at an early age, so kind of a very close to home fascination as well as very far from home one.

Q. Do your clients request pollinator-friendly (wildlife habitat) garden design or do you suggest it?

I try to help my clients be aware of the bigger picture. A pollinator garden is fine, but it's great to get people excited about both plants and insects, to truly think about the whole property as a pollinator garden. Americans in general see the garden as an object versus an environment. People love to visit gardens in England and in Europe. They may not realize it, but they are drawn to them because they are total environments: the landscape is integrated, and that's what makes those gardens so compelling.

Q. Is the trend toward using native plants a recent development?

It started in the late 1800s when people learned to identify them. Since then it has been cyclical with proliferations in the 1920s, then the 1960s, and it has grown from there. The three keys have been education, exposure, and availability. Through native plant conferences, people have learned to grow native plants. More importantly, we now have nurseries that are propagating these plants and not collecting them from the wild.

Q. What really is native? What about nativars?

There are many definitions, but you have to look at it from an ecological perspective. Plants should function as part of an ecological community. A plant can be native to large parts of the country or to a very specific region. If I were to only grow plants truly, specifically native to my area, it would severely limit what I could grow.

I don't like the term nativar, but it's for marketing, I understand. They are cultivars; it doesn't matter if it's a native or an exotic; they are all still cultivars. People assume it's a hybrid, but far and away most of our cultivars of native plants are simply selections for outstanding characteristics. There are issues with cloning, if you plant nothing but one cultivar.

Q. How does plant community restoration differ from conventional landscape planning?

In a perfect world, provenance is important and you want seed sources close to home. You certainly want to match the plants very closely to their environment, so they are exploiting an environmental niche similar or identical to where they will go. You are trying to set up a population that is self-perpetuating, able to reproduce and disperse seeds. They export and become sources of new plants, establish new populations, and enrich new communities.

Q. Best tip for designing with native plants?

Set realistic goals for yourself. What is really possible given your expertise or designer's expertise, your budget and time to maintain the garden? Do you want your garden to in some way enrich the environment? Plant lot line to lot line, and use many native plants to define them. People don't care about having an insect garden, but they want birds. Choose plants that offer the most to the broadest number of insects because they are going to bring in the birds. Then stop spraying everything to death, stop overfertilizing, and it's amazing; this stuff kind of takes care if itself.

Q. How important is art to native landscapes?

I think it's absolutely essential in many ways. Especially with meadow gardens, people find them weedy looking because there is so much fine texture. Art is a great way to give your eye something to focus on, and then you see the cornflowers and the subtle little purple prairie clovers and whatever is in the garden. I'm out in the middle of the woods, and I have these red and turquoise bistro chairs around plus a lot of sculptures. I think the garden has to delight you first or the garden will eventually go away from benign neglect because it doesn't make your heart sing.

Q. What's in your garden?

I grow a lot of early blooming shrubs and perennials. Back to those snowdrops, sometimes in January but definitely in February and early March, they are covered with bees. I love bloodroot and hepatica for early bloom. My property has a lot of native witch hazel, *Hamamelis virginiana*, still blooming for us in early December. One of the last things to bloom in October and November is aromatic aster, *Aster oblongifolius*, and it's covered with buckeye butterflies.

birds do find it an open buffet for worms (hopefully not contaminated with chemicals, too). It provides a home for grubs and attracts tunneling moles, but otherwise, pollinators look upon it as a vast, open, and therefore, dangerous expanse they must travel through to reach fragments of more worthy habitat.

REDUCE, REPLACE, AND RETHINK THE LAWN

For gardeners always looking for more places to plant, they need look no further than their lawns. It's easy to chip away at the sod, little by little expanding your flowerbeds. Or go about it more systematically and consider widening all the beds by a foot or two, or more. Or take out grass between isolated beds and combine them for even bigger areas to plant. Shrink your lawn by adding a vegetable garden, a rain garden, a water feature, or a wildflower meadow. There are so many more interesting ways to fill up space than with a monotonous monoculture of grass. Get in the car and head to the nursery!

Lots of lawns struggle in shady areas under trees and in the shadow of buildings. People sow and sow again, only to be met with frustration, even trying shade-tolerant varieties with only a mixture of luck. If you never step on this straggly turf except to mow it, consider removing it altogether and replacing it with an attractive groundcover. There are plenty of low-growing, spreading plants that have appealing foliage and even pretty flowers (with pollen and nectar) that will happily fill in the area. You'll wonder why you didn't do it sooner.

You may want to take out some of the lawn and put in a path, patio, or deck. It's not a living solution but it may make your garden more livable. Instead of concrete, use permeable pavers or stepping stones with crevice plants in between to increase water absorption and prevent runoff.

Reduce lawn area to create deeper planting beds. This means less time mowing and higher value habitat for pollinators.

LOW- TO NO-MOW LAWNS

As more people look to reduce or eliminate labor-intensive lawns, researchers and turf breeders have produced new varieties that need less mowing and are revisiting old varieties that could be used for an innovative approach. Most of the mixes use combinations of fine fescues that are also somewhat shade tolerant. Some of them grow at a slower rate; some grow tall and then flop over to create a wavy ocean-like effect. Taller grasses and lawn substitutes with less mowing disturbance provide cover and nesting sites for some pollinators. Some native varieties even have limited success if grown on the right site. For example, buffalo grass is a good alternative throughout the Plains and into parts of Texas for large areas that don't need pristine lawns. And Prairie June Grass is a possibility for up north in Minnesota, but has its drawbacks as well. Seed supplies for alternative lawn grasses are often in small supply, which makes them expensive and hard to find.

All of these low- to no-mow varieties are intended to be lower maintenance but all need care and planning at the time of installation to gain the desired results. There are some eco-minded grass seed mixes marketed nationally that may grow lush in some parts of the country but fail to fulfill their promise in others. Check with local Extension agencies for regionally appropriate low-mow lawn choices before spending the money and effort. Also check on local ordinances regarding landscapes; some ban front yard grasses above a certain height. They are usually aimed at preventing derelict properties, but can be used against homeowners who defy convention as well. You may need to inform and educate the neighbors of your plans and sometimes a small, good-humored sign will go far in alleviating any fears.

RESPONSIBLE LAWN PRACTICES

If you love your lawn like it is and don't plan to decrease its footprint, there are still ways you can reduce the amount of inputs needed to keep it looking good. These are great suggestions whether you are installing a new lawn or giving your existing lawn a second look.

- Get a soil test so you know exactly which nutrients might be lacking and you are only using what's really needed to maintain its health and appearance.
- Don't over fertilize: just because a little is good doesn't mean more is better. You'll cause excess growth that needs more mowing or even worse, burn your lawn.
- Don't overwater, as that will also cause excess growth. Use a smart timer on your irrigation system. Many lawns are growing on compacted soil created by repeated mowing, which makes it harder for water to penetrate the surface. This causes runoff, especially with overwatering. Overwatering also stresses trees growing within the area.
- Mow grass at a height of 3½ to 4 inches so that the taller blades shade the roots to lessen stress. A thicker stand of grass can help keep out competing weeds.

- Recycle your lawn clippings to reduce yard waste and give your lawn a low-dose fertilizing every time you mow.
- Consider organic fertilizers, such as corn gluten, that also help to control weed growth. Only fertilize two times a year, in spring and fall.
- Electric mowers do have some environmental costs, but avoid the noise and air pollution of gas-powered equipment. A push mower is good exercise!

EVERYTHING COMING UP CLOVER

Grass seed used to come with a bit of clover seed as well. Turns out it was a good idea. If you'd like to lower inputs to your lawn while increasing pollinator value, simply add clover to the mix. You can create an entire lawn of clover or you can overseed clover into your existing lawn. The best clover for pollinator-friendly lawns is *Trifolium repens*, or Dutch white clover since it stays low to the ground at around 3 to 5 inches tall and can blend into the grass. Recently developed miniature and micro clover cultivars grow even lower and make a great low input lawn; they can even be used for playing fields, but don't provide as many flowers for bees like white clover.

The benefits of a clover lawn are many. It is inexpensive to sow and reseeds itself to a great extent, depending upon mowing frequency. Yet it doesn't need as much mowing. Once established, it's drought tolerant with deep roots that absorb water to prevent runoff. It's also somewhat shade tolerant, growing better under trees than conventional grass. Clover has a nice soft, spongy feel to it. The densely matted surface helps to keep out weeds. It fixes nitrogen that fertilizes grass growing with it. Dog urine doesn't faze it. It may be anecdotal observation but it seems to distract rabbits from other plants. Even better, it smells good, and the pretty flowers are beloved by bees!

TOP: Clovers can be used to great advantage for cover crops, bee lawns, and bee pastures.

BOTTOM: Carefree ajugas and sedums are an attractive substitute for high-maintenance turf grass.

BEE LAWNS

Adding clover to the lawn is a good beginning for the making of a bee lawn. University of Minnesota researchers are working to find the right ingredients to make viable commercial bee lawn mixtures. They are looking at combinations of traditional cool season grasses and low-growing plants that support bees and other pollinators while meeting other criteria. As an acceptable lawn substitute it should look like a traditional lawn from a distance and be able to function as a lawn tolerating some mowing and foot traffic. It should provide the soil stability of turf grass. It should also be comprised of regionally appropriate plants that flower no taller than 3 to 6 inches high so that flower buds won't be mowed off. These plants should represent a wide diversity of types so that it has season-long food to offer bees. In northern regions, they must be cold hardy.

Until these are commercially available, people are welcome to experiment with their own mixtures specific to their region and pollinators. Choose from fine fescues, such as sheep fescue, chewing's fescue, creeping red fescue, and hard fescue. These cool season grasses are preferred since they green up sooner in the season. A number of cultivars of low-growing thymes and sedums are among the top plants for bee lawns, offering the right growth habit and favorite sources of nectar and pollen. *Prunella vulgaris*, or self-heal, and *Lotus corniculatus*, or bird's-foot trefoil, are possibilities among a wide variety of ground-hugging perennials. Naturalize spring-blooming bulbs, such as squill and crocus, into the lawn for early sources of nectar.

See where a bee lawn might fit into your landscape. Slopes that present problems for mowers are great candidates. Bee lawns near vegetable gardens and under fruit trees boost pollinator populations and therefore fruit set, resulting in bigger, better formed yields.

Don't be tempted to just stop mowing and see what comes up. It usually results in weeds like plantain, knotweed, and crabgrass—none of which are desirable for lawn areas—showing up instead.

WEEDS

Weeds may be plants, but the concept of weeds is a human invention. You've probably heard the sayings, "A weed is a plant whose virtues are yet to be discovered," or, "A weed is a plant growing in the wrong place." People go on to say that weeds are simply unloved flowers or flowers in disguise. Who decides to privilege one plant over another? There are noxious weeds and obnoxious weeds; often they are one and the same. In some cases though, a weed is in the eye of the beholder.

Butterfly weed, ironweed, jewelweed, rosinweed, milkweed, sneezeweed, and milkweed beg to differ. Libeled or mislabeled as weeds, these native plants are treasured for their beautiful flowers as well as rich habitat value.

Weeds are looked upon as chaotic, wild, uncivilized, and coarse. For gardeners, weeds can be defined as plants that interfere with the enjoyment of their garden, the aesthetics, the production, or habitat value of the landscape. Many of the plants considered weeds are non-native, man-made problems, brought over from other countries in ship ballast, pant cuffs, or bags of crop seed when America was being settled and plowed. These weeds do best on what is called disturbed soil. They thrive on farm fields, vacant lots, railroad sidings, abandoned industrial sites, and in our gardens.

Weeds can be classified as simply annoying and in other cases categorized as invasive. Invasive plants are recognized as being a threat to the environment for displacing native plants and causing further consequences to humans and wildlife. Purple loosestrife is a common example; the beautiful purple blooms belie its domineering growth habit that chokes wetlands. Like many invasive weeds, it is a prolific self-seeder; one plant produces up to 2.7 million seeds each year.

Some weeds make people miserable. Poison ivy and wild parsnip amp it up by harming people directly, causing horrible blisters and burning rashes upon contact. Mile-a-minute, aptly named for its

FLOWERING LAWN SUBSTITUTES

Blue moneywort, *Lindernia grandiflora:* Zones 7–10. Ground-hugging light green foliage with blue flowers in summer. Moderate spreader, 1 to 2 inch tall, grows in full to part sun in all soil types. Attracts bees and butterflies. Tolerates moderate foot traffic.

Cheddar pinks, *Dianthus gratianopolitanus 'Petite':* Zones 4–9. Blue-gray foliage with clove-scented pink flowers in mid-spring. Moderate spreader, 1 to 2 inches tall, grows in part to full sun in all soil types. Attracts bees, butterflies, and hummingbirds. Tolerates moderate foot traffic.

Corsican mint, *Mentha requienii:* Zones 6–10. Carpet-like green foliage with tiny blue-purple blooms in summer. Moderate spreader, only $1/_8$ to $1/_4$ inches tall, grows in most soil types including moist. Attracts butterflies. Tolerates moderate foot traffic.

Creeping red thyme, *Thymus praecox,* **'Coccineus':** Zones 4–9. Carpet-like green foliage with small red-purple flower spikes in summer. Moderate spreader, 1 to 2 inches tall, grows in full sun in most soil types with good drainage. Attracts bees, butterflies, and hummingbirds. Tolerates moderate foot traffic.

Double bird's foot trefoil, *Lotus plenus:* Zones 4–10. Densely matted dark green foliage with ruffled yellow blooms in summer. Fast spreader, 1 to 2 inch tall, grows in full to part sun in most soil types. Attracts bees, butterflies, and hummingbirds. Tolerates moderate foot traffic. Considered invasive in some areas.

Dwarf bugleweed, *Ajuga* **'Chocolate Chips':** Zones 3–9. Bronzy, green foliage, blue blooms in mid-spring. Moderate spreader, 1 to 4 inches tall, grows in part shade to sun, most soil types. Attracts bees, butterflies. Tolerates moderate foot traffic.

New Zealand brass buttons, *Leptinella squalida:* Zones 5–9. Gray-green fringed foliage with yellow blooms in summer. Moderate spreader, 1 to 2 inches tall, grows in part to full shade in all soil types. Attracts butterflies. Tolerates moderate foot traffic.

Ornamental strawberry, *Fragaria chiloensis:* Zones 4–9. Scalloped, shiny green leaves with white blossoms in late spring to summer. Moderate spreader, 1 to 4 inches tall, grows in full to part sun in all soil types. Attracts bees, butterflies, and hummingbirds. Tolerates moderate foot traffic.

Pussytoes, *Antennaria carpatica:* Zones 2–8. Dense mats of gray foliage with white flower clusters in summer. Slow spreader, 1 to 4 inches tall, grows in part to full sun in sandy soils with good drainage. Attracts butterflies. Tolerates light foot traffic.

White stonecrop, *Sedum album,* **'Coral Carpet':** Zones 3–10. Fleshy dark green foliage turns red in heat or cold, with white blooms in summer. Moderate spreader, 1 to 2 inches tall, grows in part shade to part sun to full sun in most soil types with good drainage. Attracts bees, butterflies, and hummingbirds. Tolerates moderate foot traffic.

People see weeds. Pollinators see food.

rampant growth rate, is also called tear-thumb for the fish hook-shaped thorns that draw the blood of anyone who tries to remove it.

Weeds that are highly destructive, aggressive, and extremely difficult to remove, to the point of being harmful to the environment, are put on official noxious weeds lists. Labeling certain weeds as noxious is done to attempt to keep them from contaminating commercial seed mixes. It is illegal for seed mixes to contain over a certain percentage of these pernicious plants. It also marks them out for when their removal from private property is demanded by law.

Even though they may be the bane of many gardeners' existence, such horrible scenarios make little dandelions look pretty tame. Conditioned for so long to see them as lawn invaders, lately people are coming around to their better attributes. Not everyone may want them in their salads—although the bitter greens are indeed edible, prized in the old days for their reputation as a system-cleaning spring tonic. For bees they are a valuable source of nectar in the first days of spring when other nectar-bearing blooms are still scarce. That's a good case for looking the other way when they pop up in your grass. They look cheerful after a long winter. At least let them flower before you deal with them.

Anyone remember Euell Gibbons, the popular naturalist who coined the phrase, "Eat your weeds?" Common garden weeds, such as lamb's quarters, or *Chenopodium album*; wild purslane, or *Portulaca oleracea*; and stinging nettles, or *Urtica dioica*, are edible and can be cooked and used like spinach. In fact, these plants contain high amounts of vitamins, minerals, and desirable omega-3 fats. Stinging nettles are hard to harvest, but no longer sting after they are cooked. Be sure to correctly identify any plant you plan to eat. Some species of butterflies use them as host plants for their larvae. Beneficial insects such as parasitic wasps also find the flowers of these weeds attractive. With all these features in the plus column, you might want to take the term further and call them beneficial weeds.

There are other situations when the boundaries are even blurrier. For example, creeping bellflower (*Campanula rapunculoides*) and creeping Charlie (*Glechoma hederacae*) may bloom in shades of blue-purple, but their merits are mired somewhere in shades of gray. Commonly found growing with wild abandon in many gardens, they are both considered invasive plants. Creeping bellflower produces a showy flower spike and grows from a deep taproot with little connective threads, while creeping Charlie ambles along the ground expanding from nimble root nodes to form a solid mat. They both defy complete removal. Yet bees love the flowers of these ravaging plants. What's the right way to proceed?

When pollinators seek food, they aren't choosy. With a dearth of flowers in the landscape, they don't see plants in terms of invasive or not. However, creeping bellflower crowds out other plants, including native choices that could be successfully grown in that area. If you plan to undertake its removal, be sure to eradicate a majority of the plant so that it doesn't bully new plants all over again. You should also plan to replace it with plants that offer just as much nectar and pollen to bees. When it comes to creeping Charlie, the phrase "pick your battles" comes to mind. Creeping Charlie often grows in shady areas where nothing else will. While invasive, it is not considered an ecological threat to healthy native plant communities. It is actually an effective groundcover in certain situations, and lately some experts are giving it a pass due its bee-friendly nature. Unless you can find a flowering replacement of equal habitat value, it might be advisable to make peace with Charlie.

When it comes to classifying weeds, check with your local extension and state department of natural resources for noxious weed lists for the demon weeds in your specific region. One area's noxious weed—such as English ivy in the Pacific Northwest—is simply another area's groundcover. Much of this depends upon whether the region's growing conditions favor that species to too much advantage.

WORTHY WEEDS

While not recommended for intentional planting, members of this list deserve a second look before you hack them down with a hoe. Not noxious, but sometimes annoying, these common weeds are worthwhile to bees, flower flies, butterflies, and other pollinators. Consider leaving them be on the wild edges, far corners, or unused portions of your property for readily available nectar and pollen sources. If concerned about their spread, mow or trim back after blooming before they set and disperse seed. Note that some of these so-called weeds appear on other plant lists as cover crops, garden perennials, or lawn alternatives, reminding everyone that the definition of what constitutes a weed is still a matter of attitude and placement.

NAME	SCIENTIFIC NAME
Chickweed	Stellaria media
Chicory	Cichorium intybus
Dandelion	Taraxacum officinale
Great mullein	Verbascum thapsus
Ground ivy (creeping Charlie)	Glechoma hederacae
Henbit	Lamium amplexicaule
Milkweed	Asclepias syriaca
Pennycress	Thlaspi arvense
Prostrate spurge	Chamaesyce maculata
Purslane	Portulaca oleracea
Red clover	Trifolium pratense
Shepherd's purse	Capsella bursa-pastoris
Wild mustard	Sinapis arvensis
Yellow sweet clover	Melilotus officinalis

Still use caution with so-called vigorous plants not yet considered invasive; some haven't yet reached the critical mass needed for the leap. Beware: these plants are often gifted by neighbors and sold at plant sales due to their ease of growing and enthusiastic nature.

Decide which weeds have the potential to become problems in your garden and which ones have pollinator value while being relatively harmless. Learn to identify problematic weed seedlings so they can be eradicated way before they become a bigger issue. The best time to weed is right after a good soaking rain when the ground is soft. Non-native weeds often green up faster in spring, giving away their least-favored status. At this stage, they can be pulled out or scraped off with a hoe. Know which method is best for that particular weed; some weeds multiply faster when you don't get the entire root and regenerate from even the tiniest sliver left in the ground.

WILDFLOWER MEADOWS

Meadows seem like such peaceful places; bees hum while butterflies dance above delicate wildflowers and swaying grasses. So it's surprising to learn meadows are created by natural calamities when droughts or floods wipe out trees and vegetation. Prairies differ from meadows in that they arise in areas that don't favor the growth of trees; they're created and in many cases still maintained by fire. Ancient meadows were originally formed by volcanic eruptions and glacial activity. In the wild, meadows eventually give way to the surrounding forests. Nowadays, modern meadows are more likely to develop on unused agricultural grounds.

So what's with those meadow-in-a-can seed mixes; is it really possible to make a meadow right out your front door without some sort of cataclysmic event?

Establishing a wildflower meadow requires careful planning and preparation prior to installation.

Well, it's not quite that easy, but yes: a meadow can be within reach. Manmade wildflower meadows are a picturesque, plant-rich alternative to the labor-intensive, manicured lawns that cover so many suburban and rural expanses. They are wonderful for attracting and supporting large numbers of pollinators, birds, and other small wildlife. Meadows installed in open areas of parks and schoolyards make great outdoor classrooms where students can observe plant and insect interactions. Meadow-like plantings can even be scaled down to fit smaller properties. And better yet, they can be created without a natural disaster; however they require lots of careful planning, planting, and maintenance, or the results will veer close to disastrous. Once they are established though, they do become beautiful, almost self-regulating ecosystems with only need for occasional human intervention.

Detailed instructions for meadow plantings are available on some Extension websites appropriate for each region. Reputable wildflower seed sources also share practical "in the field" knowledge on their retail websites and packaging. Beware any source that promises instant results. Avoid using novelty wildflower seed packets or giveaway packets that can contain high percentages of weedy plants. Meadows are not made in a day. Before you toss out the first seed there are things you'll need to consider:

- Make sure the light conditions can support a meadow. The site should be in an open, sunny area capable of sustaining strong growth of both wildflowers and grasses. Edges may be in shadow at times, but optimum sun exposure for most of the area should be at least eight hours a day.
- Measure the area of the proposed meadow to determine the amount of plant material or seeds needed for the project as well as any other potential costs for labor or equipment.
- Decide how existing vegetation is to be removed. Removing lawn with a sod cutter creates a ready-made planting bed. However, weedy vegetation may require repeated mowing and tilling in addition to possible herbicide applications before the area is ready for planting.
- Study the topography of the area to see whether it's flat or sloping, wet or dry; learn the drainage patterns to better handle erosion issues. South-facing slopes will be hotter and favor certain species. North-facing slopes are not optimal sites for meadows.
- Get a soil test done to determine soil pH and characteristics. Know whether you are dealing with clay, loam, or sand soils to help choose the right plant selections and preparation options for the site.
- Note the presence of buildings, fence lines, and utilities where meadow plantings should allow a buffer zone for mowing. This is also crucial if using controlled burns instead of mowing for management.
- Determine if there is adequate accessibility for equipment needed for installation and maintenance.
- Make note of surrounding properties and any vegetation that could encroach upon the meadow and plan for mowed buffers to control undesirable weeds. Notify neighbors of your plans and the benefits of meadow plantings, and keep them posted on your progress.

Armed with this information, you'll be better equipped for the fun part: choosing your meadow plants. Seeds are the most economical method, but for smaller properties, you may want to purchase small plugs to hasten the process or to bolster the appearance of seed-sowed plots in their early days. Most of you will want to buy one of the commercial seed mixes of wildflower and grass species specifically marketed for your region and growing conditions. There are also wildflower seed sources that sell individual species so you can, after doing your homework, create a custom mix for your particular site. Some folks purchase extra amounts of certain flower seeds to "paint" broad swaths of bolder colors within the meadow.

Site preparation is the most important part of the meadow planting process and determines the whole success of the meadow. The removal of existing vegetation may take an entire season and test your patience. Yet it's vital to have a relatively weed-free, smooth, and friable seedbed when starting out. Furthermore, planting in fall allows time for the seed to undergo "cold stratification" (winter cold and wet), which aids germination.

For sites less than an acre, it's recommended you broadcast the seed by hand. Add an inert material, such as sand, sawdust, or peat moss, to the bucket of seeds to help spread it evenly. Since meadows are made of tightly knit plant communities, a thick sowing mimics that, helping your meadow plants keep out weeds. Cross back and forth to ensure even coverage, then rake the seeds lightly to cover. Use a roller to press seeds into the soil to make sure they have good contact. Hopefully, rain will do most of the watering, but if dry conditions are expected, a light watering is required. Put down a light layer of straw to mulch to hold seeds in place.

For large areas over an acre, seed planters, drills, and even hydro-seeders are the recommended planting method. If you are inexperienced with this kind of machinery, look into hiring professional help. If you decide to undertake the work yourself, the best course of action is to first contact your county Extension for advice and safety precautions.

Just when your seedlings emerge and you're eager to see flowers, the meadow planting should be mowed according to regional recommendations on the seed mix. Contrary to practical thinking, this gives the plants a chance to develop deep roots and helps the meadow plants create a thick mat to block wind-blown weed seeds from germinating. By the second year, you'll finally start to see the magical effects of the meadow, drifts of wispy blooms and feathery grasses, butterflies and buzzing bees along with birds and small scampering animals. From then, rotational spring mowing is advised to help the ecosystem along. Leave plants standing in fall to catch snow to insulate the roots. The plant debris will provide cover and overwintering sites to a number of pollinators and other beneficial creatures.

While not necessary, some people employ controlled burns such as those done on prairies to regenerate their meadows every few seasons. Fire helps to remove built up plant debris and return nutrients to the soil. Where burns aren't practical, mowing and raking can substitute for the fire. Even those familiar with the practice should only do controlled burns with appropriate permits and support.

Established meadows are able to sustain themselves without additional watering except in extreme drought conditions. No fertilizer is needed. Weeding is not advised unless noxious weeds appear. Then it's recommended you spot treat with herbicide if absolutely needed rather than pulling them, which turns up the soil and exposes more weed seed to the sunlight and water needed to germinate, perpetuating the cycle.

BEE PASTURES

Not exactly a wildflower meadow but more complex than a field of cover crops, bee pastures provide bees with good nutrition that in turn fosters future generations of bees. The idea isn't new but has gained traction as pollinator populations decline due to simple lack of food as a consequence of dwindling sources of forage. Even though you can't corral them, honeybees are sometimes viewed as livestock, so it only makes sense to pasture them in a similar fashion, especially when they're used for large-scale commercial pollination. Wild bees and other pollinators also benefit from these plantings too.

Bee pastures usually comprise around half an acre but can be scaled down to fit the home garden in much the same way as when planting cover crops. Seeded in the fall, single-year pastures are composed of bee-friendly annuals, such as clovers and wildflowers. This type of pasture may be planted to bloom in coordination with an adjacent orchard or other crop of interest. The close proximity of the bee pasture to that crop takes into account the shorter flight range of various native bee species over long-haul honeybees. Fruit and nut growers are especially keen to attract blue orchard bees, considered a more efficient alternative to the beleaguered honeybee.

A bee pasture of this type should have at least three different flowering species so that blooms are available for the intended time period. They are quick and inexpensive to sow but require reseeding every year. Bee pasture seed blends have recently found their way to the market and can be ordered online, usually a mixture of clovers, cornflowers, alyssum, phacelia, and bluebells, according to region. If bulk seed orders are too big for your home garden space, think about sharing with other like-minded gardeners.

Multi-year pastures include more perennials, with some woody vines and shrubs. They demand more maintenance with weeding and pruning. Perennials will need dividing every three or four years to remain vigorous. However, these pastures have the advantage of a wider range of plant material as well as richer sources of nectar. Finally, permanent pastures include trees, more woody plants, and perennials. This level of diversity ensures that there are never gaps in bloom availability.

Creating bee pastures with an unbroken succession of blooms entices more pollinators to visit and extends the foraging season for many, inducing them to nest nearby. Along with the bee plants, the pasture should have areas to suit both types of nesting styles for native bees, with bare ground and mud for ground-nesting and tree debris for cavity-nesting. When food is abundant in the early spring months, they're encouraged to stay and nest, building up beneficial bee populations. It's important that these floral havens are located well away from farm fields that could possibly be sprayed with insecticides.

COVER CROPS

These hard-working, multi-tasking crops cover all the bases. As a living mulch, they stabilize soil and reduce splash and erosion while also suppressing weeds. Their roots can break up soil as well as bring minerals and nutrients closer to the surface. Once tilled under, they improve soil quality by adding organic matter to the soil. Yet they are sometimes planted simply for aesthetic value. Certain cover crops provide rich pollinator habitat. Although cover crops are a cornerstone of sustainable agriculture, they can be used for the same benefits in the smaller home garden. Just think of your vegetable bed as a miniature version of the farmer's field.

Cover crops fall into two categories: legumes, such as beans, peas, vetch, and clovers; and grasses, such as annual rye, oats, rapeseed (canola), winter wheat, and buckwheat. Some cover crops are also referred to as "green manures," specifically those that are meant to be turned under to improve soil fertility. The legumes also provide another layer of fertility since they "fix" nitrogen. They grow in cooperation with certain soil-dwelling bacteria that live on the root nodules of the legume. These bacteria take nitrogen gas from the air and convert it for the plant's use. When the legumes die and decompose, the residual nitrogen remains in the soil and becomes available to other plants. Legumes are used this way in crop rotation to recharge the soil and reduce need for synthetic fertilizer.

Annual cool weather crops are those planted in fall and tilled into the soil come spring while perennial crops are used for grazing, erosion control, and aesthetics. Cover crops can be a part of a succession planting scheme, used for example after an early maturing crop, such as lettuce, keeping the field somewhat productive until it's tilled under in the fall. For preparing a bed for the next season, they can be planted late in fall up to four weeks before a killing frost so they grow, go through winter, and put on an early flush of growth in spring before being turned under.

Cover crops specifically for pollinators fall somewhere in between. Most cover crops are usually mowed or tilled before flowering; however, more recently some are allowed to bloom to be used as habitat to attract beneficial insects and pollinators to farm fields, orchards, and vineyards for natural pest control and to boost pollination. They can do the same for your own edible gardens. For the purpose of luring pollinators, vetch, clover, flax, phacelia, and mustards are the best crops. Your choice of cover crop will be dependent upon hardiness zone, flowering schedule, and pollinators you hope to attract. The flower shape can also be customized to influence the type of bee you want to attract for pollinating a specific crop. You might want to use red and crimson clovers, for example, to draw in bumblebees to pollinate tomatoes. Open flower shapes, such as canola, mustard, and buckwheat will attract a wide variety of bees.

PLANT LISTS

HOW TO USE PLANT LISTS

Plants included on these lists are suitable for a broad range of conditions and are widely available at garden centers and mail order nurseries. There are hundreds of other suitable plants for pollinator-friendly gardens—too many to list in one place. Consider these suggestions a good place to start and a great leaping off point as well. Verify your hardiness zone before purchasing plants; often there are more cultivars within one species that will fit your zone.

For plants specific to your region, especially those for coastal and desert regions, check out the pollinator-friendly planting guides at www.pollinator.org. Simply enter your zip code to find more plants for your specific ecoregion.

LANDSCAPE PLANTS FOR BEES		
ANNUALS		
These old-fashioned and beloved flowers provide food for bees filling in the floral gap when other plants are blooming. They are easy to start from seed and inexpensive to purchase as plants.		
COMMON NAME	**SCIENTIFIC NAME**	**COMMENTS**
Alyssum	*Lobularia maritima*	
Bachelor buttons	*Centaurea cyanus*	
Calendula	*Calendula officinalis*	
Cosmos	*Cosmos*	
Marigold	*Tagetes*	Note: Native and non-native designations for annuals are omitted because of their endless variety; they are too numerous and varied to easily designate.
Moss rose	*Portulaca*	
Snapdragon	*Antirrhinum*	
Sunflower	*Helianthus*	
Sweet William	*Dianthus barbatus*	
Zinnia	*Zinnia elegans*	

Bees find common marigolds quite interesting. Daffodils (pictured) and other bulb flowers are a much-needed source of nectar and pollen in early spring.

BULBS AND CORMS

Many of these flowers bloom in early spring when much-needed nectar is scarce. They are an important addition to the bloom sequence of a pollinator-friendly garden.

COMMON NAME	SCIENTIFIC NAME	COMMENTS (N) NATIVE / (NN) NON-NATIVE
Allium	*Allium*	(N) and (NN)
Anemone	*Anemone blanda*	(NN)
Bluebells	*Mertensia virginica*	(N)
Blue camass	*Camassia*	(NN)
Crocus	*Crocus*	(NN)
Daffodil	*Narcissus*	(NN)
Fritillaria	*Fritillaria*	(NN)
Glory-of-the-snow	*Chionodoxa*	(NN)
Grape hyacinth	*Muscari*	(NN)
Iris	*Iris*	(N) and (NN)
Lily-of-the-Nile	*Agapanthus*	(NN)
Siberian squill	*Scilla siberica*	(NN)
Spanish bluebells	*Hyacinthoides hispanica*	(NN)
Snowdrops	*Galanthus*	(NN)
Winter aconite	*Eranthis hyemalis*	(NN)

SHRUBS

Shrubs aren't often thought of as bee plants, but many offer the flower-cluster bloom structure preferred by bees. They are an excellent source of pollen and nectar for native bees.

COMMON NAME	SCIENTIFIC NAME	COMMENTS (N) NATIVE / (NN) NON-NATIVE
Azalea	*Rhododendron*	(N) (NN)
California lilac	*Ceanothus*	spp. and cultivars
Chokeberry	*Aronia*	(NN)
Common lilac	*Syringa*	(NN)
Golden currant	*Ribes aureum*	(N)
Holly	*Ilex*	(N) and (NN)
New Jersey tea	*Ceanothus americanus*	(N)
Oregon grape-holly	*Mahonia aquifolium*	(N)
Pussy willow	*Salix discolor*	(N)
St. John's wort	*Hypericum*	(NN)
Viburnum	*Viburnum*	(N)
Witch hazel	*Hamamelis*	(N) and cultivars

TREES

*As part of the spring nectar flow, many blooming trees provide
a significant source of pollen and nectar to bees.*

COMMON NAME	SCIENTIFIC NAME	COMMENTS (N) NATIVE / (NN) NON-NATIVE
Basswood or Linden	*Tilia Americana*	(N)
Black cherry	*Prunus serotina*	(N)
Catalpa	*Catalpa*	(N) and (NN)
Crabapple	*Malus*	(N) and (NN)
Maple	*Acer*	(N) and (NN)
Mountain ash	*Sorbus*	(N) and (NN)
Redbud	*Cercis canadensis*	(N)
Serviceberry	*Amelanchier*	(NN)
Sourwood	*Oxydendrum arboreum*	(N)
Tulip tree	*Liriodendron tulipifera*	(N)

LEFT: A honeybee dives into the funnel–shaped flower of an azalea.

RIGHT: Depending upon the cultivar, witch hazel, or *Hamamelis*, blooms in late fall or early spring when other food is scarce.

ZEALOUS FOR ZINNIAS

A single packet of seeds gives you so much butterfly bang for your buck. It seems that butterflies never met a zinnia they didn't like, yet they do play favorites with some over others. Most people report more frequent visits to taller varieties, but there are exceptions. Accessible nectar with a comfy landing pad makes them such a popular flower. Grab your camera because often they are so busy nectaring they forget that you're there.

Zinnia, 'Lilliput': This 1870s heirloom received significantly more visits from more species in a University of Kentucky research study. It should be featured in every butterfly garden.

Zinnia, 'Zowie': This new variety draws them like a moth to the yellow-orange-pink flame. This one just keeps blooming and bringing in the butterflies all summer.

Zinnia, 'State Fair': Another old-timey variety that grows very tall and blooms in a wide assortment of colors.

Zinnia, 'Oklahoma': Even with the double bloom structure, this variety is a magnet for butterflies. It comes in a good selection of bold colors and is somewhat mildew resistant.

Zinnia, 'Cut and Come Again': An heirloom cutting flower in bright candy colors. You'll have to share them with the butterflies. Keeps producing until season's end as long as you keep cutting.

Zinnia, 'Berry Basket': Dense petals and delicious berry colors are the hallmark of this variety. Butterflies flock to these florist-quality blooms.

Zinnia, 'Blue Point': This award-winning zinnia has dahlia-like petals in a broad mix of clear, bright colors.

Zinnia, 'Benary's Giant': Giant dahlia-like blooms top tall stems in shades of pink, wine, and red all they way to lime. The breathtaking blooms can sometimes reach six inches across.

You can't beat old-fashioned zinnias for bringing pollinators to your garden.

LANDSCAPE NECTAR PLANTS FOR BUTTERFLIES

 ANNUALS

A broad spectrum of blooms appeal to butterflies. Plant these in wide swaths or massed in beds for the best butterfly activity.

COMMON NAME	SCIENTIFIC NAME	COMMENTS
Brazilian verbena	*Verbena bonariensis*	
Cosmos	*Cosmos*	
Dahlia	*Dahlia*	
Floss flower	*Ageratum*	
Globe amaranth	*Gomphrena*	
Heliotrope	*Heliotrope*	*Note: Native and non-native designations for annuals are omitted because of their endless variety; they are too numerous and varied to easily designate.*
Impatiens	*Impatiens*	
Lantana	*Lantana*	
Marigold	*Tagetes*	
Mexican sunflower	*Tithonia*	
Pentas	*Pentas*	
Phlox	*Phlox drummondii*	

LEFT: Butterflies can't resist Brazilian verbena, *Verbena bonariensis.*

MIDDLE: Sulphur butterflies are often found fluttering around *Impatiens.*

RIGHT: *Lantana* is an inexpensive and widely available pollinator plant.

SHRUBS

Lots of butterfly garden plans fail to acknowledge the role shrubs can play in butterfly habitat.
Flowering shrubs offer a consistent source of blooms year in and year out.
Count on them as part of the backbone of your butterfly garden.

COMMON NAME	SCIENTIFIC NAME	COMMENTS (N) NATIVE / (NN) NON-NATIVE
Azalea	*Rhododendron*	(NN)
Bluebeard	*Caryopteris*	(NN)
Bottle brush	*Callistemon*	(NN)
Common lilac	*Syringa*	(N)
Manzanita	*Arctostaphylos*	(N)
Mountain laurel	*Kalmia*	(N)
New Jersey tea	*Ceanothus americanus*	(N)
Cinquefoil	*Potentilla*	(NN)
Rhododendron	*Rhododendron*	(N) and (NN)
Seven-son-flower	*Heptacodium*	(NN)
Spirea	*Spiraea*	(NN)
Sweet pepperbush	*Clethra*	(N)
Sweetspire	*Itea*	(N) and (NN)

A fritillary butterfly enjoys the spectacular blooms of the bottlebrush shrub, *Callistemon*. Bees and hummers love them, too.

TREES

Trees are incredibly important as larval host plants for hundreds of species of butterflies. However, some offer nectar as well and shouldn't be forgotten for their contribution.

COMMON NAME	SCIENTIFIC NAME	COMMENTS (N) NATIVE / (NN) NON-NATIVE
Crabapple	*Malus*	(N) and (NN)
Crape myrtle	*Lagerstroemia*	(NN)
Red twig dogwood	*Cornus sericea*	(N)
Mimosa	*Albizia*	(NN)
Pear	*Pyrus*	(N)
Purple leaf plum	*Prunus cerasifera*	(NN)
Redbud	*Cercis canadensis*	(N)
Tulip tree	*Liriodendron tulipifera*	(N)

VINES

These vines add multi-level structure to the garden while supporting butterflies.

COMMON NAME	SCIENTIFIC NAME	COMMENTS (N) NATIVE / (NN) NON-NATIVE
American wisteria	*Wisteria frutescens*	(N)
Cross vine	*Bignonia capreolata*	(N)
Honeysuckle	*Lonicera*	(N) and (NN)
Passion flower	*Passiflora incarnata*	(N)
Trumpet vine	*Campsis radicans*	(N)
Virginia creeper	*Parthenocissus quinquefolia*	(N)
Virgin's bower	*Clematis virginiana*	(N)

American wisteria is a beautiful rest stop for butterflies. *Shutterstock*

LANDSCAPE PLANTS FOR HUMMINGBIRDS

ANNUALS

For some lucky folks, some of these flowers may be grown as perennials in warmer climates. They are all guaranteed to please the hummingbirds.

COMMON NAME	SCIENTIFIC NAME	COMMENTS
Agastache	*Agastache*	
Canna lily	*Canna × generalis*	
Flowering maple	*Abutilon*	
Four o'clock	*Mirabilis*	
Fuchsia	*Fuchsia*	*Note: Native and non-native designations for annuals are omitted because of their endless variety; they are too numerous and varied to easily designate.*
Geranium	*Pelargonium*	
Flowering tobacco	*Nicotiana*	
Impatiens	*Impatiens*	
Jewelweed	*Impatiens capensis*	
Little cigar	*Cuphea*	
Nasturtium	*Tropaeolum*	
Penstemon	*Penstemon*	

Red *Nicotiana* is sure to attract hummingbirds.

🐦 VINES

These beautiful vines bloom in mid- to late summer when hummingbird populations are at their largest and hungriest. They add a great vertical element to your garden while helping hummers to tank up for their long migration.

COMMON NAME	SCIENTIFIC NAME	COMMENTS (N) NATIVE / (NN) NON-NATIVE
ANNUAL		
Cardinal climber	*Ipomea × multifida*	(N) and (NN)
Chilean glory vine	*Eccremocarpus scaber*	(NN)
Cypress vine	*Ipomea quamaclit*	(N)
Scarlet runner bean	*Phaseolus coccineus*	(NN)
Small red morning glory	*Ipomea coccinea*	(N)
Spanish flag	*Mina lobata*	(NN)
PERENNIAL		
Coral honeysuckle	*Lonicera sempervirens*	(N)
Trumpet creeper	*Campsis radicans*	(N)

Hummingbird favorite cardinal climber, *Ipomea × multifida* is a vigorous bloomer that's easy to grow from seed.

TREES AND SHRUBS

Sometimes you see hummers at unexpected places, such as these blooming trees and shrubs. Add flowering trees to enhance the pollinator habitat value of your total landscape.

COMMON NAME	SCIENTIFIC NAME	COMMENTS (N) NATIVE / (NN) NON-NATIVE
Azalea	*Rhododendron*	(NN)
Blueberry	*Vaccinium*	(Cultivar)
Bottle brush	*Callistemon*	(NN)
Flowering quince	*Chaenomeles*	(NN)
Lilac	*Syringa*	(NN)
Mimosa	*Albizia*	(NN)
Red buckeye	*Aesculus pavia*	(N)
Rose of Sharon	*Hibiscus syriacus*	(NN)

ALL SORTS OF SALVIAS

Some of these salvias may be harder to find, but they are all guaranteed to bring in the hummers with their striking blooms. Hummers tend to prefer taller varieties to shorter bedding types. They are considered annuals or tender perennials in many areas.

COMMON NAME	SCIENTIFIC NAME	COMMENTS (N) NATIVE / (NN) NON-NATIVE
Anise-scented sage	*S. guaranitica 'Black and Blue'*	(NN)
Autumn sage	*S. greggii*	(N)
Bog sage	*S. uliginosa*	(NN)
Darcy's Mexican sage	*S. Darcyi*	(N-Mexico)
Eyelash sage	*S. blepharophylla*	(NN)
Giant Brazilian sage	*S. subrotunda*	(NN)
Mountain sage	*S. microphylla, 'Hot Lips'*	(N-Mexico)
Pineapple sage	*S. elegans (N-Mexico)*	(NN)
Roseleaf sage	*S. involucrata*	(NN)
Scarlet sage	*S. splendens*	(NN)
Texas sage	*S. coccinea*	(N)

Salvias are in big demand by many pollinators, bees, butterflies, and hummingbirds alike.

Butterfly weed, *Asclepias tuberosa* 'Hello Yellow', is one milkweed that monarch butterflies use as a larval host plant.

CHAPTER 4

SUSTAINING BUTTERFLIES
WITH LARVAL HOST PLANTS

A CLOSE LOOK AT CATERPILLARS

Has something been devouring your dill or chomping
on your hollyhocks? Congratulations, you might already
be a butterfly host! What many people think is a pesky
"worm" may really be a welcome caterpillar doing what
caterpillars do before they turn into butterflies, which is
eat. However, they don't just eat any plant; the specific
plants they require to survive are called *larval host plants*.
Turns out that supporting pollinators on their journey
through complete metamorphosis requires way more than
just planting pretty flowers.

You might say this is the not-so-sexy part of butterfly
gardening, yet that's exactly what it is. You see, after
butterflies mate, like lots of folks, they need proper
childcare. Offering food and drink will encourage fully-
fledged butterflies to visit your gardens. But if you want
them to do more than visit occasionally, if you want them
to stay around and increase their populations, you have to
provide food for their children.

It all starts when you see a butterfly fluttering back and
forth over a certain area of your garden where others might
be feeding upon flowers. That's a male butterfly looking
for a mate using a behavior called *patrolling*. Some species
will spend time instead perching on a tall plant or structure
while searching for a suitable mate. Either way, butterflies
don't have great sight so once he spies a comely candidate,
he'll fly in for a closer inspection.

A black swallowtail caterpillar munches
on parsley, or *Petroselinum*.

TOP: The adult monarch can nectar on many different flowers but its larvae is a specialist that needs milkweed to survive.

BOTTOM: Larvae of the painted lady are more generalists; they prefer thistles but feed on asters and mallows as well.

Once he's found a female, and one of the right species, he'll fly above or behind her to let her know he's interested, releasing pheromones while exhibiting a courtship behavior of excited fluttering. He hopes she's amenable, but this isn't a given. If already mated, she may rebuff his advances with body language that says "Not today, buddy." If she agrees, though, she'll find a place to land where they can join abdomens and mate. It's not unusual for butterflies to continue mating in flight.

Once mated, the female has to find a place to lay her fertilized eggs, and that isn't just any dead leaf or dusty corner. She seeks out the unique and necessary food plants that her species has co-evolved with over countless generations, so that her young will be able to start feeding immediately after they hatch, thus aiding their survival. The female looks for the right leaf shape and shade of green, then ventures closer. Butterflies taste with their feet (how cool is that!), so she drums her feet on the leaf to make sure it's the right plant. Then she lays one, several, or lots of eggs; under the leaf, on the leaf, on the flower or stalk, or perhaps at the leaf axils, depending upon her species.

Some butterfly species lay clusters of eggs, the resulting caterpillars feeding in hungry groups. Although there's safety in numbers, these groups can attract the attention of predators. The female monarch uses another strategy, usually laying a single egg on each milkweed plant so as not to draw attention to the new larva while lessening competition for food.

When it comes to larval host plants, butterflies are considered either specialists or generalists in their food needs. The monarch butterfly is a specialist (also called an *obligate*) in that although the adult butterflies can nectar on a number of flower species, its larvae can *only* eat milkweed (*Asclepias*) as a host plant. Other butterflies, such as the painted lady and mourning cloak, utilize numerous species of larval host plants: they are the generalists.

Once the eggs hatch, the caterpillars go through several stages, usually four incarnations, shedding their skins while growing larger each time in forms called *instars*. The whole process of molting is kind of like teenagers going through bigger and bigger sneakers! And just like those teenagers, they are ravenous eaters for a time.

With their soft bodies and slow, creeping mode of locomotion, caterpillars are easy prey for birds, wasps, other insects, and animals. To protect themselves during this vulnerable phase of life, caterpillars use various defense mechanisms. Simply blending in to the scenery being green is a common strategy. Other may hide inside a curled up leaf. Some only feed at night. This may explain why it's so hard to find caterpillars in your garden. It turns out they are hiding in plain sight.

Some caterpillars emit horrible smells to fend off predators. Other caterpillars hope to appear fearsome, sporting an enormous eyespot in hopes they seem much larger than they really are, to scare off attack. You have to admire the swallowtail caterpillars that adopt the clever guise of a bird dropping, then take it one step further in the last instar stage changing once more, this time to look like a little snake.

While in most cases the best thing you can do is provide lots of host plants for these hungry fellows, some gardeners worry their landscapes will look nibbled upon and gnarly—if not totally stripped of foliage. It should be said that the non-native cabbage white butterfly can inflict great damage to agricultural crops. However, with most butterflies, the damage, if you choose to call it that, (the larvae just call it dinner) is usually minimal. Healthy plants can tolerate having a portion of their leaves chewed, and eventually regrow more. After all, one reason plants exist is to feed animals: their beauty is just a wonderful byproduct. Besides, what good is a picture-perfect garden if there are no butterflies in that picture?

Once the caterpillar has finished eating and reached the last instar stage, the next step is to form a chrysalis that seemingly transforms into a butterfly through magic. The caterpillar spins a silk pad with a connecting strand and attaches it to a sturdy leaf or stem. Often this isn't the host plant but a nearby plant. The caterpillar goes through a number of contortions as it almost turns itself wrong side out shedding skin and creating the shell that will hold its burgeoning adult body. Depending upon the species, it can takes weeks or months before the butterfly emerges. On average, butterflies only live around three weeks as the fully formed butterflies you love to watch float over your garden.

GETTING STARTED WITH LARVAL HOST PLANTS

When you look up "larval host plants for North America" online, chances are you'll be presented with lengthy lists of plants from areas all over the country. On average, you'll find there are at least 100 species of butterflies near your home with even more in the warmer West, southern Texas, and into Florida. That number decreases as you move north to Canada and over toward New England.

You can easily feel overwhelmed deciding where to start.

First, you'll want to narrow it down to the region where you live. Remember these guys have been eating local long before the rest of us discovered the idea. Go out into your garden and simply observe. It's best to begin by identifying a few common butterfly species, maybe three to five that you see regularly. Then zero in on the host plants they prefer. Refer to the lists provided (page 80) to see which larval host plants butterflies are utilizing around your garden right now. You can use this information to decide whether to plant more of those varieties or where to add new ones.

Note: If you do only one thing to help out butterflies, consider planting just three additions to your garden: dill, hollyhocks, and milkweed (for more on milkweed, see page 83). These appealing plants support a large number of a wide variety of species: you might even say they are the holy trinity of host plants. And the more you plant, the more butterflies you'll see.

While it might be fun to plant one of everything on the larval host plant list, your garden will be more successful to larval survival if you plant larger quantities of fewer varieties. It's easier for butterflies to find larger plantings of what they need among all the greenery they survey. Having lots of the same plants in one area also reduces the energy the female has to expend as she lays eggs. And once the eggs hatch, there's plenty of food to go around.

Hollyhocks along with dill and milkweed are some of the best larval host plants for your garden.

Just as pollinators have adapted to native plants for nectar sources, they have also evolved to using particular native plants as larval hosts. Native plants support the highest numbers of pollinators as a food source for their young. But wait a minute: what about the dill just mentioned? Didn't it originate on another continent?

Dill is a member of the carrot family, *Apiaceae*. Before dill and other domesticated members of this plant family, such as parsley, fennel, and anise, were brought to this country, black swallowtail caterpillars ate other wild members of the carrot family, such as golden Alexander, cow parsnip, and prairie parsley. When butterflies resort to using non-native members of the same plant family, the plants are called *adopted exotics* or *introduced plants.* This is one of the reasons herb gardens can be good for pollinators.

This phenomenon doesn't happen often or within every plant family; in fact, many times only a sole member of a plant family is host to a particular butterfly, and life can be precarious for these uber-specialists. Most of these pollinator-plant synergies have evolved over thousands of years, yet there are instances of pollinators coming to prefer non-native sources of larval host plants, like one area on the West Coast where conservationists and butterfly enthusiasts are tussling over whether non-native fennel should be eradicated in some areas in spite of Anise Swallowtails heartily devouring it.

LOOK TWICE AT THE LABEL

No matter where you buy seeds or plants, in person or through online or mail-order sources, gardeners agree that those brief blurbs on labels, packets, and catalogs can be confusing when deciding what to plant.

In response to gardeners seeking more information about pollinator plants as well as jumping on this new marketing message, plant breeders and sellers now often include cute little butterfly icons or exclamations, such as "Attracts Butterflies!," on product descriptions. While a great beginning, this is almost always directed at the nectar-producing aspect of the plant. Until the habitat value of host plants becomes more common knowledge and hits the plant marketers' radar, you'll have to do your own homework for larval host plants, starting with the lists in this chapter, perhaps.

THE LARVAL HOST PLANT LIST

Picture a butterfly garden, and a patch of beautiful flowers comes to mind, with butterflies browsing and sipping nectar from brightly colored blooms. But when asked to imagine a larval host garden for butterflies, what springs to your vision? Purposely planting for an army of hungry, hungry caterpillars? It's not a familiar concept to many gardeners.

Instead of a flowerbed here or there, many of these plants are the stuff of meadows, pastures, or forests. So how do you fit them into a home landscape? Actually, larval host plants can fit into the bigger picture of your yard—you probably have some already and don't know it! Top to bottom, these hosts cover a wide array of diverse plant types: trees, shrubs, grasses, herbaceous perennials, cover crops, herbs, even vegetables and all the way down to the violets that invade your lawn. Consider the possibilities!

You may be surprised to see the list of trees included here. Yes, trees! Most people don't consider that forest and woodlands play such a vital role in butterfly survival. Amazingly enough, the oak tree in your front yard has the capability to support around 500 Lepidoptera (butterflies and moths) species while the blooming perennials in your flowerbed, although important, too, nourish numbers in the tens.

Worried that a particular host proves too popular and leaves your flowerbed with a gaping hole or unsightly chewed-upon plants? If so, landscapers suggest placing these plants in out-of-the-way places (but not too far from forage plants) or at the back of borders where their appearance is not as important.

This list contains plants that are widely available to grow for a number of common and widespread species of butterflies. Except for the specialist butterflies, certain families of butterflies tend to utilize one or more related plant families. For example, black swallowtails use members of the wild carrot family, which includes lots of everyday herb plants, such as parsley and dill; lots of sulphur butterflies are connected to the legume family; skippers use numerous grasses; and various admirals, among others, feed on tree species. (However, buckeye butterflies are all over the map with monkey flowers, snapdragons, and more.)

Follow the list as a leaping off point for your larval host plant quests. If you want to support a more obscure butterfly or an endangered species in your area, you'll want to drill down into more detailed information. See Resources (page 170).

TOP: People are surprised to learn that trees such as this mighty oak, *Quercus*, support numerous butterfly species.

BOTTOM: Queen Anne's lace, *Daucus carota*, is favored as a host plant by a number of swallowtail species.

LARVAL HOST PLANTS

HERBS	
NAME	**SCIENTIFIC NAME**
Anise	*Pimpinella*
Dill	*Anethum*
Fennel	*Foeniculum*
Mint	*Mentha*
Parsley	*Petroselinum*

PERENNIALS	
NAME	**SCIENTIFIC NAME**
Aster	*Aster*
Hardy hibiscus	*Hibiscus*
Hollyhock	*Alcea*
Mallow	*Althea*
Milkweed	*Asclepias*
Pearly everlasting	*Anaphalis*
Purple coneflower	*Echinacea*
Turtlehead	*Chelone*
Wild indigo	*Baptisia*
Wild senna	*Senna*

TOP: Wild indigo, *Baptisia australis*, attracts numerous butterflies as both nectar and host plant.

MIDDLE: Asters are not only important nectar plants but significant host plants, too.

BOTTOM: Caterpillars of the common buckeye use hoary vervain, *Verbena stricta*.

COVER CROPS	
NAME	**SCIENTIFIC NAME**
Alfalfa	*Medicago*
Clover	*Trifolium*
Mustard	*Brassica*
Sweet Clover	*Melilotus*
Vetch	*Vicia*

"WEEDS"	
NAME	**SCIENTIFIC NAME**
Nettle	*Urtica*
Pepperweed	*Lepidium*
Plantain	*Plantago*
Thistle	*Cirsium*
Violet	*Viola*

GRASSES	
NAME	**SCIENTIFIC NAME**
Big bluestem	*Andropogon*
Blue gramagrass	*Bouteloua gracilis*
Little bluestem	*Schizachyrium*
Pennsylvania sedge	*Carex pensylvanica*
Prairie dropseed	*Sporobolus heterolepsis*
Switch grass	*Panicum*

Once common along roadsides, vetch, *Securigera varia*, is a larval host to many butterfly species.

TOP: Willow, *Salix*, is host to a number of butterflies including the viceroy and mourning cloak.

BOTTOM: Some shrubs, such as this highbush cranberry, *Viburnum tilobum,* are excellent hosts as well.

TREES	
NAME	**SCIENTIFIC NAME**
Aspen	*Populus*
Black cherry	*Prunus*
Cottonwood	*Populus*
Elm	*Ulmus*
Hackberry	*Celtis*
Hawthorn	*Crataegus*
Oak	*Quercus*
Poplar	*Populus*
Tulip tree	*Liriodendron*
Willow	*Salix*

VINES	
NAME	**SCIENTIFIC NAME**
Honeysuckle	*Lonicera*
Hops	*Humulus*
Moonflower	*Ipomea*
Passionflower	*Passiflora*
Pipevine	*Aristolochia*

SHRUBS	
NAME	**SCIENTIFIC NAME**
False indigo	*Amorpha*
Pawpaw	*Asimina*
Spicebush	*Lindera*
Viburnum	*Viburnum*
Wild lilac	*Ceanothus*

WHAT'S SO SPECIAL ABOUT MILKWEED?

Milkweed is undergoing a marketing makeover. Lately, it's gone from weed to wonderful. Once you make the connection between milkweed and monarch, it's easy to see that this plant is a superpower in the world of pollinators.

There are all kinds of milkweed across the country, and their names (such as sand, prairie, clay, and swamp) give a clue to where they live or what they look like (short, tall, green, purple, whorled, and more). The diversity of milkweeds is mirrored in the huge number of pollinators that utilize this plant for survival. It also provides food and shelter to a sizable community of birds, spiders, beetles, beneficial insects, even frogs and other small animals.

However, the monarch butterfly's interdependence goes bigger: the majestic butterfly is a specialist in that milkweed is the only host plant capable of sustaining its larvae. No milkweed, no monarchs. It's as simple as that. So specialists like the monarch are at much greater risk when habitat destruction and fragmentation decimate the one and only larval host plant they depend upon.

Milkweed isn't only vital to monarchs for food during the larval stage of its life cycle. It performs another crucial task for both the caterpillar and the adult butterfly in creating a defense mechanism. As the larva eats the leaves, it ingests toxic chemicals called *cardenolides* found in the milky sap. These horrible-tasting compounds make monarchs unpalatable to birds and other predators. Birds come to associate the color and pattern of monarchs with a terrible taste and potential tummy ache, and learn to avoid them at all costs. A few other butterflies mimic the monarch's appearance in order gain this protection, too.

First disappearing as the country was settled with farms and cities and eventually suburban sprawl, you might say milkweed lived on the edge. For some time, milkweed was plentiful along farm fields and roadsides, the narrow strips left to nature acting like links in a chain connecting thousands of similar small habitats.

Forty years ago, farm kids were paid to pull up milkweed in crops and adjoining property. It was considered a scourge on the land with its enthusiastic ability to spread by windblown seed or creeping rhizome. No one yet realized the delicate connection between the bright orange butterfly and this rough and tough plant. In fact, it was only just around 40 years ago that scientists finally confirmed monarchs do indeed migrate. Until then people suspected they left, but weren't sure where they ended up.

As farming and road building practices "improved," a misguided effort to achieve tidier fields and roadsides meant large-scale mowing took down more milkweed. And although Lady Bird Johnson lobbied for wildflower conservation back in the 1960s, pretty wildflower species more popular than milkweed ended up beautifying the back roads and interstates because of her efforts. When given a choice of poetically labeled posies, such as Indian paintbrush, bluebonnets, and winecups, who wants something with "weed" in its name? Even then, milkweed was in sore need of a good publicist!

Look closely at common milkweed, *Asclepias syriaca*, to discover a monarch caterpillar.

ASK THE EXPERT:

DR. KAREN OBERHAUSER

Karen Oberhauser is a professor in the Department of Fisheries, Wildlife, and Conservation Biology at the University of Minnesota, where she and her students conduct research on several aspects of monarch butterfly ecology. Her research depends on traditional lab and field techniques, as well as the contributions of a variety of audiences through Citizen Science. In 1996, she and graduate student Michelle Prysby started a nationwide Citizen Science project called the Monarch Larva Monitoring Project, which continues to engage hundreds of volunteers throughout North America. Karen is passionate about the conservation of the world's biodiversity, and believes that the connections her projects promote between monarchs, humans, and the natural world promote meaningful conservation action. In 2013, Karen received a White House Champion of Change award for her work with Citizen Science.

Q. Why monarchs? Did you choose them or did they choose you?

I chose them. I was a first year graduate student looking for a research question. I was interested in parental investment in offspring and wanted to study a species where I could really quantify that in males and females. I was thinking of birds because male birds help feed their young. Someone suggested butterflies since the males transfer this package of nutrients to the female while they're mating. So I started working on reproductive ecology in monarchs and then started my own students and followed their interest. I have now studied many, many different aspects of monarch biology and about halfway through I got very involved in conservation.

Q. Why are people so fascinated by monarchs?

You know, many biologists or researchers are used to studying something that nobody really cares about, and monarchs are certainly not the case. It's humbling to study something that people care so much about. They're beautiful, they're very familiar, and that's partly connected with their beauty. They're also widespread and occur in places that are pristine prairie habitat but also people's city gardens. They're interesting, a lot of features of their biology are fascinating like the interaction between plants and animals and the toxicity of milkweed that protects the monarchs. Finally, they're just impressive, that an organism that weighs as much as a paper clip can migrate thousands of miles to overwintering sites that they've never been to before. Monarchs are a flagship species that people use to make a connection to nature.

Q. What are some of the greatest challenges that monarchs face?

The number one problem is habitat loss. As a migratory species they need a series of habitats available to them at the right time and right places. But once the population gets small, and the monarch population now is really small, then they become more vulnerable to all the other threats. There's climate change, so we not only have to save habitat that they need now but we need to think about what will be available for them in 50 years. Lots and lots of pesticides like mosquito sprays and neonicotinoids that are being used to control pests have impacts on non-target insects like monarchs, but also herbicides that allow people to spray after crops emerge from the ground. This causes habitat loss because there used to be milkweed growing in those fields and that's no longer there.

Q. Are there still unsolved mysteries about monarchs?

The biggest and coolest mystery is fine scale migration. We know how they fly south, but we don't know how they find those very specific overwintering sites. That would trump all to know how they navigate to those same places year after year.

Q. What's in your yard?

I'm looking out the window and yes, it's very easy for people to find my house, it has a prairie. Unfortunately I'm surrounded by monocultures of Kentucky bluegrass. I have many species of milkweed, meadow blazing star, lots of asters, stiff goldenrod, coneflowers, and lots of baptisia; let's see what else do we have...

As people became more aware of the ecological importance of native plants along roadways, "no-mow" and "no-spray" signs popped up signifying small efforts at reclaiming this valuable habitat. But the success of these undertakings is dwarfed by the destruction wrought by new and more powerful agricultural innovations and herbicides that recently came on the market.

With the advent of genetically modified crops (GMOs), yield increased while sometimes allowing for actual decreased use of pesticides, but these advances in the area of production sounded a death knell for any milkweed that still found a spot left to grow. Now when crops such as corn, cotton, soybeans, canola, and alfalfa are genetically embedded with the ability to resist herbicide damage, fields can be sprayed with herbicides to kill weeds that might have made it before. Even the tough, leathery leaves of many milkweeds are no match. Efforts are made to reserve acreage for buffer zones, but often herbicide drift negatively affects those areas, too.

As milkweed continues to vanish from these rural settings and routes, it may be up to home gardeners to make a difference and make up the difference in milkweed plantings.

GROWING MILKWEED

Many people are afraid to plant milkweed, worried that it will take over their yards—while this isn't really the case. The list on 88 notes a number of milkweed species and cultivars suitable for home landscapes with information about the best sites and conditions for their particular characteristics. *Asclepias tuberosa* (butterfly weed) and *Asclepias incarnata* (swamp milkweed) are the most popular and adaptable perennial types for conventional

A large stand of common milkweed, *Asclepias syriaca*, may be better suited to casual landscapes, such as this charming farmhouse.

garden beds. Whether started by seed or plant, be sure to find a permanent place, since once mature the large taproot makes it difficult to transplant.

Asclepias curassavica (tropical milkweed), with its striking flowers, is a favorite for annual beds and colorful containers. The South American native can get a bit weedy when it grows happily in warmer climates. Recently, some monarch experts were warning people not to plant tropical milkweed in moderate climates where it will persist into winter. According to reports, a plentiful supply of this milkweed tempts large numbers of monarchs to breed and overwinter in place instead of continuing their migration. If a cold snap comes along in these areas, it has the potential to wipe out roosting monarchs. If this is the case in your region, tropical milkweed should be treated as an annual and removed at summer's end.

Less refined types of milkweed are appropriate for more naturalistic settings or larger spaces where they can spread without resentment. It's not anywhere near as rampant as bamboo or mint, but certain types will spread through rhizomes underground. Even so, they can be used in more manicured settings with a few considerations on management. Stick with *Asclepias tuberosa* if you want a non-rhizome spreading type.

One tactic is to allot a dedicated area or bed for milkweed only. You'll be won over with this island of milkweed when not only monarchs but also a variety of other butterflies, plus a diversity of bees and hummingbirds, visit it. If you're worried about what the neighbors will think, space the milkweed so the planting looks more intentional and add a piece of garden art. Better yet, add a birdbath or bee block (see manmade habitat section on page 137) for these collateral creatures. In addition to the beautiful blooms, many milkweed flowers are very fragrant. Some varieties have a pleasing vanilla scent.

If reseeding is a worry, and milkweed does reseed, there are ways to stop or redirect this part of the life cycle, although the milkweed once again is just trying to ensure survival of the species. Some gardeners will simply cut off the fuzzy or bumpy seedpods as they appear, depending upon which type, before they mature, pop them open, and fling their contents to undesirable locations.

If you want seeds to form but want to decide where they are sowed, seal the pods shut with rubber bands or twine. Some folks secure them with netting from old onion bags. When the pods are completely dry, you can harvest the seeds to plant or share. Milkweed seeds need to be pre-chilled before sowing to simulate natural conditions. If starting indoors, sow in pots, preferably biodegradable pots so as not to upset the growing taproot, cover with plastic wrap and refrigerate for 21 days. Remove from fridge eight to ten weeks before planting. If sown outdoors in either fall or spring, simply cover with a light dusting of soil and water until germinated.

Alternatively use the attractive pods in dried flower arrangements and other crafty projects.

Few pests bother milkweed; however, aphids can be troublesome depending upon your tolerance. The little squishy, yellow, oval creatures will gather in huddled masses at the top parts of the plant. If the infestation is not bad, the sucking damage will be minimal and other insects and birds will make a meal of them. If their numbers increase, you can pick or brush them off, but if that makes you squeamish, hose them off or spray with soapy water. The same can be done if whiteflies, scale, or spider mites show up and cause concern.

A MILKWEED FOR EVERY GARDEN

Milkweed may be the only larval host plant for the monarch butterfly, but lucky for gardeners (and monarchs), there are many types of milkweed. There's sure to be a milkweed that suits the particular climate, conditions, and style of your garden while supporting the beautiful butterfly. The following list is roughly in order of usefulness to common gardeners.

Asclepias tuberosa, Butterfly weed: Hardy Zones 3–9, summer blooming, grows in average, well-drained soil but tolerates dry soil and drought once established. Beautiful orange flowers sit in clusters above the lance-shaped leaves. At under 3 feet, it's good in borders, butterfly gardens, and prairie plantings, planted in drifts or spotted throughout. Can be used to naturalize larger areas. Easy to grow from seed or plant, locate carefully since the taproot makes it difficult to transplant. Named cultivars with yellow flowers include 'Hello Yellow' and 'Western Gold'.

Asclepias syriaca, Common milkweed: Hardy Zones 3–9, summer blooming, grows in average soil, including shallow, rocky soil, and tolerates drought. Native to the eastern US. Pink to mauve pompon-shaped flowers bloom on the upper part of plant; broad, almost leathery leaves can take a lot of damage from hungry larvae. May be aggressive for borders and groomed gardens, but does well for naturalized plantings and wilder areas of the yard.

Asclepias incarnata, Swamp milkweed: Hardy Zones 3–6, blooms mid- to late summer, grows in medium to wet soils. At 4 to 5 feet, it's taller than most milkweeds, with vanilla-scented, pink flowers in tight clusters atop lance-shaped foliage. A great candidate for low, moist spots in the garden. Ideal for streamside or next to ponds. Suitable for rain gardens and naturalized areas. Named cultivars include 'Cinderella', 'Soulmate', and white 'Ice Ballet'.

Asclepias sullivantii, Prairie milkweed: Hardy Zones 3–7, blooms early to midsummer, grows in medium to wet soil. Bright pink clusters of blooms with thick, broad leaves. Sometimes called smooth milkweed, this more refined version of common milkweed comes highly recommended for butterfly gardens, borders, and rain gardens.

Asclepias curassavica, bloodflower or tropical milkweed: Hardy Zones 9–11, when grown as an annual blooms all summer long; grows best in rich, well-drained soil with regular watering. A native of South America adapted to tropical and subtropical areas of the US. Red-orange flowers with yellow crowns float above dark green, lance-shaped leaves. An attractive addition to flower borders, butterfly gardens, and cottage-style gardens. Stunning in containers, too. 'Silky Gold' is an all-yellow cultivar.

TOP: Butterfly weed, *Asclepias tuberosa*, behaves well in conventional perennial plantings.

BOTTOM: Swamp milkweed, *Asclepias incarnata*, can tolerate wetter soils than some other milkweed varieties.

Asclepias speciosa, Showy milkweed: Hardy Zones 3–9, blooms early summer, grows in dry to medium soils, tolerates drought once established. Native to the western US and parts of the Midwest. Blooms are rosy pink to purple globes on the upper area of the plant above broad, velvety leaves. A more manageable milkweed, it's a striking addition to borders, beds, and butterfly gardens.

LESS COMMON MILKWEED SPECIES

These less common milkweeds may be only available as seeds. Although the tried and true varieties will guarantee lots of monarch visits, it's fun to try unusual ones to see the incredible variety among the members of this valuable plant family. The following list is roughly in order of usefulness to the common gardener.

Asclepias asperula, Antelope horns: Hardy Zones 7–9, unique, globe-shaped cluster of green petals with accents of purple. Native to Texas, best suited for the Southwest.

Asclepias physocarpa, Goose plant: Hardy in Zone 8 and above, distinguished by comically hairy, balloon-like seedpods. A tall milkweed grown as an annual, late blooming, a sure conversation piece in addition to monarch helper.

Asclepias exaltata, Poke milkweed: Hardy Zones 3–9, this is a shade-tolerant milkweed. It's unlike most milkweed, with loose, drooping, white flower clusters and bright green oval leaves.

Asclepias hirtella, Tall green milkweed: Hardy Zones 3–7, a taller milkweed at 4 feet. Grows in medium dry to wet soil. Rough white flowers born in clusters. Tolerates partial shade.

Asclepias virdiflora, Green comet milkweed: Hardy Zones 3–8, green starburst-shaped flowers, grows in sandy and rocky soil. Native to western US, endangered in some areas.

TOP: *Asclepias curassavica* is a tropical milkweed grown as an annual in many parts of the country.

BOTTOM: This pretty urban garden contains quite a few larval host plants, such as pagoda dogwood, baptisia, and milkweed.

Bumblebee queens overwinter in nests beneath grasses such as this beautiful 'Little Bluestem', *Schizachyrium scoparium*.

CHAPTER 5

NESTING SITES FOR BEES

HOW BEES RAISE THEIR YOUNG

For those of us who first learned about bee nests from Winnie the Pooh, or perhaps Yogi Bear and Boo Boo, a little clarification may be needed. The tubby little bear had it right when he dipped his paw into the honey tree hole. Too bad for Yogi and Boo Boo though, standing beneath that tree limb with their bucket. They would have gone hungry and without honey in the real world since the cartoon shows they were actually prodding a wasp nest!

Unless you think of those nice white boxes that managed-bee populations call home, it's hard for most folks to imagine just what a bee nest looks like. This may be because they're not as noticeable to the casual observer, but also because there's a shortage of the right nesting conditions in many home gardens.

Some bees, such as the wild honeybees in the Hundred Acre Wood across the pond, nest in tree cavities. However, only 30 percent of native bees in North America nest in some form of wood, usually a stump or plant's stem. The other 70 percent are ground nesters.

Many folks are surprised to find out not all bees live in large social colonies. In fact, over 90 percent of bee species in North America are solitary bees, the females of which build nests and supply food for their young without any help. The visible part of these bees' lives that we watch with fascination is only a small portion of their existence. They spend most of their life cycle tucked away in the nest as an egg, larva, and pupa before emerging to forage and reproduce for only a matter of weeks.

TOP: The familiar white hives that house managed honeybee colonies.

BOTTOM: Some bees use tree cavities for raising their young.

Upon hatching, solitary male bees hang around nests or cruise meadows, flowerbeds, orchards, and woodland edges looking for suitable mates. Once mated, a female solitary bee is tasked with building a nest and stocking up on food for her young, usually a mixture of nectar and pollen called "bee bread" that she makes and leaves to nourish the new generation. With only a few exceptions, when this job is done, she dies.

TOP: Save trimmings from hollow-stemmed shrubs for tunnel-nesting bees.

BOTTOM: Leafcutter bees leave telltale signs of their handiwork. They line their nests with the "quilt pieces" of leaves.

Although called a solitary bee, she doesn't always nest alone: many bees may nest at a single site. Burrowing into the soil, many ground-nesting solitary bees build single nests at the same location in large groups called *aggregations*. Some species of bees may share a communal entrance hole, like an apartment lobby, while tending separate brood cells off the main shaft. Ground nests are often confused with anthills or wormholes, although the holes are a bit larger; you should look for bees' comings and goings to be sure.

Whether a particular bee prefers loose, sandy soil or clay, below ground these nests take on different shapes. Some nests branch off like fingers from the entrance hole while others form a radial pattern off a long vertical cylinder. The female bee lines the brood cells with her own waxy or oily secretions to protect the cells from rain and spoilage. Notably the polyester bee is named for the cellophane-like secretion called *linalool* she uses to line her brood cells. This material not only acts to block fungus and bacteria, it also protects the nest from normal rainfall as well as seasonal flooding, necessary since the species often nests on creek banks.

It's more accurate to call bees that nest above ground, in wood, tunnel-nesters. Many use pre-drilled sites such as abandoned beetle tunnels in stumps or dead and dying trees referred to as snags. Others chew through the pithy part of plant stems. Large carpenter bees have strong enough mandibles that they can actually chew through wood to create their nests. Some tunnel-nesting bees are opportunists, occupying crevices in rock walls, used insect nests, and other existing cavities they might find.

Within these tunnels, they then create linear nests consisting of separate brood cells. This is why they are sometimes also referred to as *tube nesters*. Most of these bees line the brood cells with some sort of waterproof material to prevent damage from water, fungal disease, and scavenging insects that could compromise the food they've left for the egg once it hatches. The brood cells are lined with secretions and often reinforced with chewed up leaves and petals, tiny pebbles, sand, sawdust, resins, or other substances, depending upon the species. When a tube is filled, the end is sealed up with the same substance.

Leafcutter bees go one better and line their cells with pieces of leaves specially cut and "stitched" like a quilt with their jaws to seal the cell. You can tell their work by the circular holes found in many leaf margins around the garden. Yet this cosmetic damage is often blamed on slugs or caterpillars!

Incredibly, many native bees have the ability to determine the sex of each egg. They position males nearest the nest opening for several reasons. Since females mate with multiple males, the males are considered more expendable. The first bees to emerge are more likely to be the subject of bird and insect attacks, so the "women and children first" rule doesn't apply. The surviving males are at the scene and ready to mate when newly hatched females leave the nest.

Some bees don't bother to build nests; they simply steal nests from other bees. Like the nest-robbing birds they are named after, cuckoo bees will "borrow" a place to raise their young. Some cuckoo bees are *cleptoparasitic*, meaning that after they break and enter into another species of solitary bee's nest, they'll lay their own egg. According to which type of cuckoo bee they are, the larva will consume the food intended for the original egg, with that egg either destroyed by the female or the larva upon hatching. Other cuckoo bees are social parasites in that they invade the nest of social bees and manage to make the residents raise their young for them.

Bumblebees are social nesters. They use abandoned rodent nests, clumps of grass, or other cavities for raising their young. They form clusters of brood cells that look like little pots made of a waxy substance. They are different from all other bees in that more than one egg is laid in each cell. Bumblebees provide food for their offspring on an as-needed basis, bringing and storing nectar for the colony. Bumblebees don't overwinter their young, so nests are annual affairs, leaving only a few mated queens at the end of the season when the rest of the colony dies off. Come spring, the new queen lays eggs that become the next generation, then she raises them and forages for them until they can start a new colony. Once that happens, she stays in the nest producing eggs.

Unlike honeybee hives, bumblebee nests may contain only a few tablespoons of honey. So why do bears raid bee nests in real life? The bears' thick fur protects them from the bees' wrath, but they still risk stings on the nose in exchange for the whole contents of the nest; a veritable picnic basket of protein-rich larvae, the brood cell material, and sometimes even the bees themselves.

ALLOWING PLACES FOR NESTING

When shopping centers and housing developments eat up open spaces in and around our towns and cities, it's obvious how wild areas for nesting bees disappear. In our gardens, the cause is more cultural expectations and a case of misdirected good intentions.

In the quest to be both responsible community members and good gardeners, people have squeezed bees out of the housing market. Tidy, frequently micromanaged yards—where every dead leaf and piece of debris is whisked away the minute it hits the ground—are simply bad for bees. The same goes for yards where mulch is applied as a top-dressing to every single square inch of soil surface.

This bumblebee will build her nest in pre-existing cavities, such as a mouse hole or under tufts of grass.

Q. How important was the connection with nature in your childhood?

When we moved to Minnesota when I was ten, suddenly the natural world cried out to me, literally, as geese flew overhead and wind rustled falling autumn leaves. The smells and textures of trees, the full measure of seasons, slowly drew me out, as did my mother's gardens.

Q. Prairie or poetry, what came first?

Poetry. When I started researching my memoir, I started reading more about plants and gardening, and that led to prairie and that led to thinking about my gardening, in different ways, local ways, and Nebraska ways.

Q. If your garden had a mission statement what would it be?

Use native prairie plants to save wildlife. Know your home ground, and know the larger world beyond your garden and what is at stake. We are losing prairie at a record pace, which leads to vanishing species, dirty water, loss in soil fertility, loss of our natural heritage.

Q. How have your garden and your gardening evolved over time?

When I first started gardening I was way too particular and worried about keeping plants where they are. Now I let my plants move around and find where they want to be. I spend a lot more time researching the ecological benefits of plants before I buy them, let alone plant them. I don't buy what's pretty at the nursery, but instead think how it will contribute to my mini-ecosystem, from soil and light conditions to wildlife.

We must unlearn what we've learned about gardening. This doesn't so much mean the appearance, but the maintenance. Gardening is not necessarily hard work if you're in tune with the place—this means the right native plant in the right place.

A S K   T H E   E X P E R T :

DR. BENJAMIN VOGT

Benjamin owns Monarch Gardens, a prairie garden consulting and design firm in Nebraska. He writes a landscape column for Houzz.com and is on the board of Wachiska Audubon Society, a prairie and wildlife conservation group that manages nearly 1,000 acres. His writing and photography have appeared in *Orion*, *The Sun*, *Nebraska Life*, several Great Plains anthologies, a Xerces Society book on butterfly gardening, and he contributed to the book *Lawn Gone*. He presents nationally on the ethics of native plant gardens in a time of climate change and extinction, as well as on low-maintenance and sustainable designs for wildlife and people. He has a Ph.D. from the University of Nebraska and lives in Lincoln.

Q. You encourage risky behavior in the garden. Please explain.

I don't mow my lawn often, maybe once a month during the summer. So we got a complaint one of the neighbors filed that said our lawn was worthless and weedy vegetation, to which I agreed! This fall we ripped out most of it. Now we have a lawn path and border going around two garden beds with native plants. There's purpose to it, it is just not a weedy mess. I was very aware that I needed to put in clumps and drifts. It's just not plant choice and placement but also what kinds of borders you have. Leaving the grass border helps us tie together with the other houses that just have lawn so we still have continuity going on. Limited to 12 to 15 species, it shows design intent, I'm trying to keep the palette simple and very clean so it doesn't appear chaotic to people.

Q. Your best design tips for native plants with pollinators in mind?

Plant thickly.

There's no reason for narrow foundation beds. Deepening the foundation beds to ten feet makes a home feel a lot more welcoming.

Let the plants teach you over the years what they want and what they provide for wildlife.

Q. What's in your yard?

I'm currently in love with round-headed bush clover, *Lespedeza capitata*. That one's good for wildlife with super unique winter seedheads. Virginia mountain mint, *Pycnanthemum virginianum*, is a freaking cool plant; the leaves smell minty yet it doesn't spread like a true mint. On a sunny day you'll see and hear bees all over it. And there are so many awesome *Liatris liatris*—short, spiky, tall, purple, white. . . .

Just as you wouldn't want a lengthy commute from your home to the grocery store when your family is hungry and waiting for dinner, neither do bees. Bees need to nest close to where they forage. Unlike other pollinators, bees may be lugging a heavy load of pollen and/or nectar between flowers and their brood cells, so they calculate the distance carefully, and so should you in your garden plans.

Flight range is a major consideration when planting flowers in relation to potential nesting sites. While honeybees travel an average of two miles between their food source and hive, native bees cover much shorter distances in their daily outings depending upon their size. Bumblebees and a few other larger bees may be able to travel a mile, but smaller species can be limited to distances around 500 feet or less. The less energy spent between forage and nest, the better for the bee.

For an average-sized yard, the distance between food and nest may not be an issue. In larger, suburban yards, there may be an opportunity for several nest sites to correspond with different forage areas. It's more about locating an area where you don't mind bees raising their young.

If you're ready to forego a few yards of mulch and expose some soil, there are a few considerations. Ground-nesting bees in general will gravitate to looser sandy or loamy soils since they are easier to excavate. Some species nest on flat ground while others prefer east- or south-facing slopes that warm first early in the day and early in the season. Deeply shaded spots aren't the best candidates. Avoid spots that tend to stay continually wet as well, although some tunnel nesting bees will seek out damp soil for sealing brood cells. Are there wilder edges around the garden or an area at the back where no one will mind occasional mud? The bare soil under evergreen trees

TOP LEFT: Nesting sites should be located within reasonable distance of forage areas.

TOP RIGHT: A patch of bare soil for bees within this blooming border isn't noticeable from the curb.

BOTTOM RIGHT: Green sweat bees dig vertical burrows in slopes or banks.

BOTTOM LEFT: Many ground-nesting bees prefer an east-facing slope.

where nothing grows may seem like wasted space but bees like the umbrella-like protection afforded by the low-hanging branches. The easiest solution may be to allow an area behind shrubs or at the back of the perennial garden to go undressed. If you're worried about aesthetics, taller perennials and shrubs can screen the holes where plants are absent. Strategically placed garden art can always draw the eye up and away from the area as well.

You might want to allow a sloping part of the yard to go "naked" as slopes are often the first choice of many ground-nesting bees. People generally don't spend time on hilly parts of the yard, so they are a great place to experiment with potential nesting sites. To allay concerns about erosion, sow patchy spots of bunch grasses with exposed soil in between to hold the ground and offer places for bumblebee nesting as well. Consider native grasses, since the roots of most extend farther into the ground than you might imagine, often deeper than the plant is tall; it's great insurance against erosion. If you want a more landscaped look, use several varieties of ornamental grasses of different heights spaced at irregular distances, still with enough bare places in between.

In summary, the best thing you can do for ground-nesting bees is…nothing! Once you've chosen or designated an area for nesting purposes, be sure to avoid heavy foot traffic or driving over it so you can avoid soil compaction. The same goes for tilling that could disrupt the life cycle in spring or fall. If ground-nesting bees happen to take up residence in the fluffy soil of your vegetable garden, resist the urge to till. In the long run, it will be better for your soil structure, since tilling disturbs the complex ecosystem of microorganisms and worms that enrich as well as aerate the soil. It is engrained in gardeners that they should tend and cultivate their patch of earth. It's harder to learn to simply leave well enough alone.

A lot of bee habitat makes it through the summer in our gardens but gets "cleaned up" come fall. Putting the garden to bed should include a thought or two for all the baby bees nestled snug in their brood cells hoping to make it through winter. It's a shame when they end up on the curb in a leaf bag full of garden debris destined for the dump.

Lots of people like to tidy up their gardens by trimming back perennials. Instead, trim back perennials and ornamental grasses at the end of the year only if they are broken or flopping over. If you can't handle the idea of leaving plants standing, make sure to leave stem stubble at a height of at least 15 inches for tunnel-nesting bees to use. Otherwise, think about leaving up the brown stalks and seedheads for a number of reasons, some of which you may know already. Left standing, the plants collect snow or leaves around their crowns, which help to insulate plant

Consider leaving perennials standing through winter to enhance habitat value.

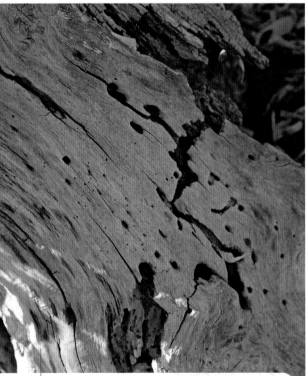

TOP: Perennials provide for winter interest and nesting sites all at the same time.

BOTTOM: Tunnel-nesting bees often use "pre-drilled" holes made by beetles for nesting.

roots concentrated near the soil surface. The seedheads provide food for hungry birds and other small animals. Beneficial bugs use them for winter condos. The seedpods offer winter interest while the garden sleeps, especially when touched with frost or daubed with snow. But now you know bee nests may be tucked away in cavities beneath these tussocks or in the hollow stems above as well. It is perhaps the best reason to leave them alone.

Whether it's spring or fall, when pruning certain shrubs and fruit bushes, you should think twice about how you dispose of the leftover trimmings. Lots of tunnel nesting bees find the pithy stems of dogwood, hydrangea, and sumac, as well as raspberry and elderberry, the perfect choice for nests. If possible, leave the pile of trimmings in a less visible or hidden area of the garden so that tunnel-nesting bees might find them, but passersby won't find them offensive. To make them more pleasing to the eye, tie them in neat bundles and stack or arrange artistically. Showing intent changes a brush pile from messy to meaningful. Call it an eco-art installation, and the neighbors won't dare question you.

With just as much flair, you can incorporate fallen branches, stumps, and logs into garden décor. Use them to anchor plantings of perennials or accent pieces among shrubs. Tunnel-nesting bees will appreciate your endeavors. You'll be surprised at the number of tiny creatures who call these woody relics home. Dead tree remains called snags are valuable for nesting sites; they are most likely already riddled with beetle tunnels and larger cavities great for a variety of nesters. They're probably most appropriate for larger properties where they aren't situated near the house though. Only leave them up if it's safe to do so. Disguise them with flowering vines if they're deemed unsightly.

Once again, sometimes the most efficient way to provide nesting sites for bees is to do nothing. Allowing for and protecting natural nesting sites, in many cases, rests on leaving property undeveloped and unimproved. This is usually easier for large suburban yards and small acreages. Smaller urban lots may not be appropriate for brush piles and tree remains. For those spaces, there are still lots of ways to support a variety of native bees with manmade habitat, such as bee blocks and bundles of nesting tubes (see chapter 7). As always, every effort counts.

WHAT TO DO WITH PROBLEM NESTING SITES

Lots of so-called problems with bees are based on a case of mistaken identity. Too many times people refer to anything that buzzes as a bee. This generic term has framed many a bee for trouble it didn't cause when it fact the culprit was some kind of wasp.

Even though experts sometimes have trouble differentiating some bee species from wasps, generally there are certain distinguishing characteristics to tell one from another. To start, wasps are shinier and more slender than bees. The old-fashioned phrase "wasp-waisted woman" comes from the curvaceous indentation between the thorax and abdomen of most wasps. As carnivores, wasps are attracted to meat, so they're always sure to attend your barbecues and picnics, especially in mid- to late summer when they are hungriest. And since they're also attracted to sweet things, they've been known to finish off that burger with a soft drink.

Bees, on the other hand, don't bother with meat; they only have eyes for the pollen and nectar, so they forage from flowers. They also tend to be fatter and fuzzy. Those hairy legs and abdomens are handy for packing on the pollen needed back at the nest or hive.

The most commonly encountered wasp is the yellowjacket. You can easily recognize them by the bright yellow bands on their tapered abdomens. And sure enough, they are often called "meat bees." They're a nuisance around public picnic areas where food is exposed and plentiful. Toward mid- to late summer (just in time for Labor Day outings), as the colony grows in size and food becomes scarce, they grow more bold and aggressive. The best tactic to avoid them spoiling a picnic in the park is to keep food covered with lids until ready to serve and then to cover it again immediately after serving.

Wasps are often mistaken for bees. Furthermore, these paper wasps are often confused with yellow jackets.

MUCH ADO ABOUT MULCH

Gardeners have gotten the message about mulch and have really taken the practice to heart. And for good reason: there are so many benefits to mulching your landscape. Mulch:

- Moderates soil temperature. Instead of keeping your plants cozy and warm like a blanket, it actually keeps the soil from undergoing extreme temperature fluctuations that can be fatal to plants.
- Adds a layer of protection to conserve loss of moisture from the soil through evaporation.
- Suppresses weeds by limiting sun exposure so that weed seeds don't readily germinate. It can also smother existing weeds.
- Helps avoid soil compaction and crusting over that makes it difficult for water to penetrate.
- When it's organic, it improves soil structure as it decomposes.
- Keeps water-born fungus and pathogens in the soil from splashing on foliage.
- Prevents erosion.
- Is an inexpensive way to cover pathways through the garden.
- Makes an attractive top-dressing that gives the landscape a finished appearance.

But enough with so much mulch! Sometimes mulch can be too much of a good thing because bees need bare soil for nesting. Added to all the other threats and challenges bees face, there's a serious shortage of plain old dirt in commercial and residential landscapes.

It gets worse, since landscape fabric and plastic are often laid under mulched areas for what's thought to be better weed proofing. Between the mulch and the extra layer, bees can't get a break, literally. Although such an underlayment sounds good in theory, the weed-proofing ability of these added layers is short lived since windblown dust and leaf

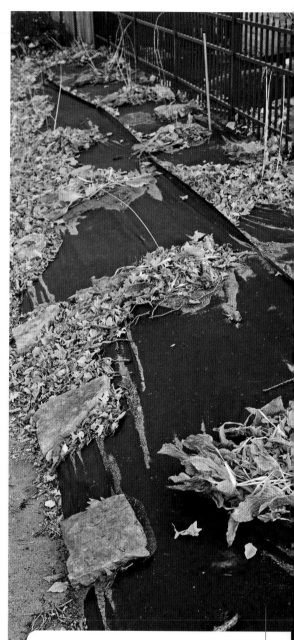

Plastic mulch is a bad deal for bees and other pollinators, as it smothers and seals off the soil.

debris collects on top, creating a thin mantle of growing medium, a comfy zone where weed seeds still enthusiastically sprout.

In addition, landscape fabric, although permeable, makes it harder for water to penetrate the surface. It also acts as a barrier to the natural breakdown of organic materials that improves soil structure. Plastic, on the other hand, completely smothers the soil, leaving a barren ecosystem underneath that excludes all the beneficial creatures that live beneath the ground and contribute to soil health. Worse yet, black plastic creates a hot, stressful environment for plants by heating the soil to extreme temperatures. Plus the planting holes poked in the plastic are insufficient to channel adequate water to reach plant roots. There are plenty of reasons to ditch the plastic.

While it's not recommended you do away with all of your mulch, consider carefully where it goes. Resist the urge to carpet your whole yard. There are lots of places where mulch makes perfect sense. These are usually areas of the garden where there is a lot of human activity and the potential for the kind of disturbance that doesn't favor peaceful places for bees to raise their young.

- Mulch is good for paths and areas that get lots of foot traffic. You wouldn't want bees nesting where you walk anyway.
- Straw mulch is great for the vegetable garden where it maintains moisture for crops. Bee nests sometimes don't survive in the veggie plot where frequent sowing, harvesting, or tilling occurs.
- Annual flowerbeds that are watered regularly and cultivated seasonally benefit from applications of fine shredded mulch. Bees would find these beds unsuitable for nests.
- Ornamental beds near entrances or play areas, where bees might spook people or pets, are good places for mulch.
- Areas that are highly visible from the sidewalk or street needed for curb appeal are good for mulch.

What does that leave? More places than you think. Sometimes the best thing is to leave parts of the yard "undeveloped" or wild. In many yards there's often a back corner or lonesome spot where no one treads. It doesn't have to take a huge amount of space. Large suburban yards and small acreages may have more opportunities to host bee nesting sites, maybe in several different areas of the property. Bee advocates suggest you aim for 50 percent of your property to be exposed soil. For more urban yards, that may not be practical, but still the goal gives you an idea of how inhospitable many home gardens have become for bees. Start small and see where it goes.

Smaller urban lots might not be able to offer natural nesting sites for bees, especially ground-nesting species, but can easily accommodate a bee bundle or block for tunnel-nesters (see Chapter 7).

Promptly dispose of uneaten food into covered trashcans. Watch your soda for wandering wasps that crawl into cans and bottles. Pour drinks into glasses or paper cups but still remain observant.

Bee and wasp nests can be a true problem if they disrupt or cancel outdoor activities or threaten people and pets. Although true bee and wasp allergies are rare, this can be another factor in deciding a nest's fate. The first course of action is to correctly identify the insect and locate the nest site. You may want to stop right there and call in the professionals if the idea of getting that close gives you a fright or risks your health. Beforehand, visit the closest university's website to get research-based information on "problems with ground-nesting insects" in your region for a better idea of the options available, in order to determine the necessity of calling in professionals. Otherwise, proceed with extreme caution and take into account the various kinds of nests you might encounter.

Wasps are social insects, meaning they live in colonies. Wasp nests start out small in the spring and usually aren't noticed until they've gained in size and population. A colony can contain anywhere from 1,500 to 15,000 wasps. Yellowjackets, depending upon the species, are both ground- and aerial nesters. If a nest is well away from your home and human/animal activity, leave it alone since they are valuable beneficial insects that prey upon lots of garden pests. And once you're near the end of the season, know that it only takes two hard frosts to shut down the nest naturally. However nests near the home, especially by doors and entrances or in the lawn where mowing can antagonize them, probably need attention.

Experts recommend treating problem nests at night when wasps aren't active. It's still important to wear protective clothing and eyewear, though. Pouring a solution of soapy water down the entrance hole of a ground nest may solve the problem. If not, use an insecticide labeled for wasp control and follow all instructions explicitly. Insecticidal dust is considered more effective than liquid because it better permeates the nest.

Some wasps, such as European paper wasps, also make hanging nests by chewing up wood and forming layers of paper-like material. With a fascinating whorled design, these nests are definitely an attractive nuisance. They can be treated with aerosol insecticides aimed at the entrance. These products shoot a stream from a distance, sometimes up to 20 feet. Wasps may sense the poison and attack, however, even at night. Nests concealed within walls and building cavities are more difficult to treat. In late fall, use of insecticides may cause the wasps to retreat inward and chew through a wall, possibly entering your home. Once again, think about hiring a professional in such a situation.

Large wasp nests found in winter or very early spring are most likely from the previous summer. Since wasps don't reuse nests, they are probably unoccupied and should be disposed of before scavenging creatures, such as carpet beetles, take over the nest and possibly get inside your home.

Another common wasp is the mud dauber. Mud daubers aren't social insects, so the female tends her nest as a single mother creating clusters of mud cells attached to ceilings and walls. She's less aggressive and relatively harmless.

Actual bee nests aren't usually reason for concern. Bumblebee nests occasionally cause problems. Solitary bees may nest in noticeable groups—you'll see them buzzing around the soil in springtime—however, they are docile and rarely sting. The easiest and least toxic way to rid them from an undesirable location is to simply turn the sprinkler on regularly. Most bees require dry conditions for their nests and will move on to another area if the soil stays damp.

DEALING WITH SWARMS

Many know it by a quaint or onomatopoetic term: rabble, hum, charm, grist, and erst. No matter the name, a swarm of bees conjures up an image of undulating energy in insect form. But there's a logical reason behind the phenomenon.

Honeybees are the only type of bee that truly swarms in large numbers. Sure, you may see a small gathering of bumblebees hovering around the outside of a nest, but those are only male bumblebees waiting to mate with emerging females. And since males don't sting, there isn't much reason to worry.

There are other times when you should be concerned about swarming bees. Africanized honeybees, sometimes referred to as killer bees, will chase and attack people when they sense a threat to their hive. European honeybees, the kinds used in commercial pollination operations around the country, usually remain pretty docile when they are in swarm mode. The problem is they are very hard to tell apart. Africanized honeybees can be found in the southern tier states from California to Georgia. Exercise extra care in these regions if you encounter a swarm.

Caution sign for visitors at a botanical garden. Swarms are usually focused on finding new digs for the colony.

When it comes to the relatively gentler honeybees found throughout the country, their swarms have nothing to do with anger. A swarm is the bee colony's method of reproduction. When weather warms and forage is plentiful, queen bees lay more eggs and as a consequence, hives can be become overcrowded. Once this happens, worker bees find it hard to keep the brood supplied with enough nectar and pollen. In this respect, swarming is a natural outcome of a strong, successful hive. More room is needed for the resulting swell in population. Beekeepers employ various methods of management to prevent their domesticated bees from swarming, but sometimes the bees have other ideas.

Within all the hive's activity, bees are communicating their exit strategy by pheromones transmitted through their food. As bees prepare to swarm, the queen bee slows laying and the older, restless bees gorge on honey as they get ready to leave the hive. When this group departs, a new queen takes over the old hive. The swarm consists of the old queen and at least half of the colony amounting from up to 5,000 to 20,000 bees. This swirling mass of bees finds an interim stopping place to cluster while scout bees look for a new home. Most people never notice a swarm since it usually finds a tree branch or other out of the way spot, although they can light anywhere such as a house, shed, or a mailbox where they do seem threatening. But while still in the swarm, which can last from a few hours up to several days, these temporarily homeless honeybees are probably still well fed and full of honey and thus not as prone to sting as they would be when guarding their hive.

At this point, it's easy for beekeepers or other professionals to collect the swarm before they have the chance to establish a new colony. They can be brushed or funneled into a container. Many beekeepers collect swarms as a side business, often performing the service for free, simply adding the extra bees to their own hives. You might even say they are "swarm chasers!"

While the swarm may be harmless in itself, it's wise to call someone to remove it in case the swarm decides to locate permanently on your property. If the swarm appears to be shrinking in size, they may already be moving in. But wait a minute: aren't we supposed to be supporting bees?

Yes, but honeybee colonies are a messy prospect without a proper hive. The scout bees look for a cavity with a 4- to 9-gallon capacity. This could be a chimney, utility box, barbecue grill, or even an abandoned car. Imagine that filled with 100 pounds of honey! They often enter cracks and holes in homes and outbuildings, finding their way into walls and attics, where it gets much harder to remove the colony.

Once the scouts find a suitable (at least for them) location for the new hive, they start to fan the site, releasing a different set of pheromones that guide the rest of the bees to it. But if they don't find a new site, the swarm will start producing comb right then and there in an attempt to establish a new home. They'll do this under the eaves of a home, in a tree, or similar spot. These exposed pop-up hives are usually short-lived due to predation by birds and other insects or inclement weather.

Once a honeybee colony makes your home its home, the trouble begins. Whether attached outside or hidden inside the structure, bees create comb, honey, and brood, a sticky combination of materials that can rot or ferment on or within the walls of your home. If the colony is left to build up without control, the damage from seepage can be substantial. If your walls are buzzing, better call someone to research the situation! Although sometimes they can kill the bees or lure the bees from the structure, removing the actual hive may require removal of plaster, drywall, and other materials. It may be next to impossible to clean out all the residue containing pheromones that could make the building attractive to subsequent swarms. This can be problematic if you live in an area where there are lots of beekeeping operations, and additional swarms are a real possibility.

Homeowners are usually careful to seal up cracks and crevices to keep out birds, squirrels, raccoons, and other critters, yet bees and other small insects can enter through the tiniest of holes. All entrance points should be sealed tightly. Expandable foam can be used to fill cavities in walls so as to make them unattractive to bees. Chimneys are difficult to bee-proof, with bees preferring the rough interiors of masonry ones that hold attached comb material better than slippery sides of prefab versions. It's recommend you call in professionals the minute you see bees hovering around chimneys and not build a fire to smoke the bees out. It may be too late: if the opening is blocked, you may have a fire risk. Professionals can install special dampers that seal the top and are effective at keeping bees as well as other animals out of your house.

Swarming may seem aggressive, but it's merely the natural result of a successful hive.
Shutterstock

Shallow depth and textured surface make this concrete leaf birdbath a perfect watering hole for pollinators.

CHAPTER 6

PROVIDING WATER
AND OTHER LIQUID NUTRIENTS

HOW BEES USE WATER

Bees get some water from flower nectar, but they still get thirsty just like humans. They need water to satisfy this thirst and to maintain their health and hive.

When bees collect water, they do it with the same efficiency as when they gather nectar and pollen. Bees search for water close to home and once they find it, fill their crops and ferry it back so it can be used for a number of things depending upon the time of year. Honeybee foragers carry water back to the hive and give it to other workers through a behavior referred to as *trophallaxis*. It may look like the bees are kissing; however, this mouth-to-mouth transfer of liquid is more complicated. In addition to unloading the water, it communicates information through various pheromones about the condition of the hive. In the process, the forager bee will learn from the worker bees whether it needs to make another water trip or not.

More water is needed during the spring, when nectar flow is not at peak, and also in the heat of summer. During the brood-rearing season, the nurse bees need lots of water to help feed their brood. They consume this water along with large amounts of nectar and pollen that they then regurgitate to make a substance called royal jelly for the newly hatched larvae to eat. As summer heats up, the worker bees will use the water to balance temperatures in the hive. They take the water and spread a thin film over the brood cells, and then vigorously fan the area to create evaporative cooling using the same mechanics as a swamp cooler to air condition the

Honeybees collect water for use in cooling the hive and feeding larvae in brood cells.

TOP: Bees actually prefer "aged water" with some algae growth.

BOTTOM: Plentiful blooms, grasses for cover, and shallow running water offer ideal habitat for pollinators.

hive. In winter, honeybees use water to dilute their stores of honey that have become crystallized in cold temperatures, re-liquefying it in order to eat it.

Like butterflies, some species of bees siphon water from puddles in order to extrude minerals and salts necessary for their health. Sweat bees will do the same but often find moist humans a perfect source for these valuable salts.

Bees seek out water wherever they can find it, and they're not too picky about its condition—in fact, they seem to prefer dirty water. It's thought that bees are attracted to water with some algae growth because of the smell. Some folks say bees like their water "well-aged." Other times, it's hard to explain why they are drawn to damp laundry or the dog's dish for hydration, other than that it was right there at hand.

Then what's the thing with bees and swimming pools? Bees seem drawn like magnets to the big blue lagoons. This is one of those cases when the intersection of humans and insects results in many complaints. There are several theories that may explain the fatal attraction. One simple reason is that when large populations of bees live near a pool, a certain number may go belly-up in the pool due to natural causes. At the height of summer when people are most likely to be sunning or swimming, bees from a hive can die at a rate of 1,000 per day. They often drop dead while out foraging, and they just happen to do so while flying back and forth above the large turquoise target.

Other theories say bees are attracted to the salts present in the pool's chemically treated water. They smell with their antennae and feet as well as their tongue, so some attribute the problem to something like "sweet foot" rather than sweet tooth. Actually, they are enticed by salt even more than sweet, so that may be why they end up in the drink. And since bees are mediocre swimmers, they end up soaked and unable to save themselves when faced with the steep sides of the pool.

If you garden with the goal of attracting bees or you maintain beehives close to your neighbor's pool, it's important to give your bees a safe alternative water source. Bees can easily drink without danger from water sources that have a graduated edge or something to grip onto. They try to stand out of the water while drinking to remain dry. So there's no problem when they drink from puddles, pond and stream edges, shallow birdbaths, and the like.

Healthy garden habitats should always include some water, even if it is just a simple birdbath. Water not only sustains wildlife, it cools the garden with its presence and elevates the landscape with light, sound, and movement. For garden ponds and fountains, floating plants, such as water lettuce, water hyacinth, duckweed, and water lilies, provide a landing pad that bees can drink from safely. A few, such as pickerelweed, water hyacinth, lilies, and lotus, are forage plants as well.

Keep your bees consistently well watered, and they will keep away from the pool. Otherwise, the pool owner may have no option but to kill bees that are ruining their right to swim unaccosted by stinging insects. If someone has a problem with bees by their pool, let them know they can spray the offending bees with soapy water, since that will kill only a small number, with the hopes of discouraging their fellow bees from visiting the pool, too. As with solving any neighborly dispute, communication is key.

POLLINATOR FAVORITES: TEN NATIVE PLANTS FOR WATER'S EDGE

What could be better for pollinators than a serving of beautiful blooming native plants with a side of water? These plants all thrive in the saturated soil found along the edges of garden ponds and naturalistic water features. Food and drink all conveniently located!

NAME	SCIENTIFIC NAME
Blue-eyed grass	*Sisyrinchium angustifolium*
Blue flag iris	*Iris versicolor*
Buttonbush	*Cephalanthus occidentalis*
Cardinal flower	*Lobelia cardinalis*
Golden Alexander	*Zizia aurea*
March marigold	*Caltha palustris*
Monkey flower	*Mimulus ringens*
Obedient plant	*Physostegia virginiana*
Pickerel weed	*Pontederia cordata*
Swamp mallow rose	*Hibiscus moscheutos*

Obedient plant, *Physostegia virginiana*, adapts well to the water's edge.

BUTTERFLIES: WHEN NECTAR IS NOT ENOUGH

While nectar and butterflies go hand-in-hand, many butterflies live on more than nectar alone and seek out non-floral nourishment from other sources. You probably couldn't conceive of the butterfly garden they really want: lots of flowers, for sure, with some mud and a side of roadkill, sticky sap, rotting fruit, and plenty of poop for good measure. It turns out that dainty, delicate butterflies engage in some downright disgusting behavior (although they don't think so, it's just another means of survival for them). Never fear, though: there are other ways to provide liquid nutrients for butterflies without turning the garden into a garbage dump! It may be as simple as a few mud puddles.

MAKE A BEE WATERER OR BEE WATERING STATION

If you want to provide safe water sources that cater to bees' particular needs, it's easy to make a simple bee waterer. With a bit more ambition you can assemble a bee watering station!

MAKE A BEE WATERER

Find a shallow saucer-shaped container a few inches deep. The saucers that accompany flowerpots are ideal, but any similarly shaped ones will do. Size can vary; you may want to make several with different diameters.

Fill the saucer with marbles or polished stones from the florist section of your local craft store. You could use clear or colored marbles depending upon the theme of your garden. There's no reason the bees shouldn't drink in style.

Fill the saucer with water and wait to see who visits. The marbles or stones allow the bees to stand on them while drinking. Refresh the water regularly to avoid mosquito problems. Be sure to locate the bee waterer near forage areas. Also take care to place it where the bees won't disturb guests or pets but you can observe their comings and goings when you want.

If you only want to enhance existing water features for bee imbibing, you can float tennis balls (the fuzzy surface is easy for bees to cling onto) or plain old sticks for landing spots. Or stay classy and float all those accumulated wine corks for the same effect.

MAKE A BEE WATERING STATION

Materials:

- One five-gallon plastic or resin flowerpot
- Small circulating water pump
- Two feet of plastic tubing
- One sheet of stucco lath
- Five short screws
- Assorted rocks and stones
- Pie pan or shallow metal dish

Turn pot upside down and trace outside opening on lath with a marker. Use leather gloves while cutting the lath into the circle. Snip five tabs into the perimeter of the lath so you can attach the lath with screws to the inside wall of the pot. Aim to have the screws fit in or under lip of the pot to avoid sharp edges. This will hold the lath across the top of the pot.

Before screwing the lath into place, install the pump at the bottom of the pot and pull the tubing and electrical cord through the lath. Screw the lath into place with sharp edges curved under so to not injure yourself and so that rocks will stay on top of it.

Fill the pot with water. Place the pie pan on the lath and fill with small stones. Add more rocks around the edges to hide the lath and give the bees additional surfaces to land on. Plug in the pump and adjust the tubing so the water lands in the pie dish. Give the bees a few days to find the bee watering station, and then have fun observing them while they drink up.

courtesy Albuquerque Beekeepers Association

Speaking of which, you may have seen large gatherings of butterflies on the side of the road or in a vacant lot that appear to be just sitting in a puddle of mud. They are doing something called "puddling." Although both genders are known to puddle, males are more likely to engage in this behavior. This is because the mud puddles attract male butterflies seeking nutrients needed to aid fertility that nectar doesn't provide. You can't help but think the puddles are the butterfly equivalent of the pub. But before you call it sexist, be assured that while not as many females actually practice puddling themselves, they do actually benefit from this behavior, too.

If you were able to magnify the scene, you'd see the male butterflies sit in damp patches of sand or mud and use their proboscis to filter feed or siphon minerals and salts from the moist earth. They in turn pass on this nourishment to the female in their sperm when they mate. Aptly called a "nuptial gift," the extra nutrients not only improve her metabolism but assist in reproduction making for healthier eggs. Puddling is an essential part of butterfly reproduction, so much so that in times of scarce rainfall or all out drought, many butterflies will go into reproductive diapause, ceasing to mate or lay eggs until environmental conditions improve.

However, sometimes a puddle is just for drinking. Butterflies can't drink from open water; they need to find a source with shallow water and a sloping edge to stay dry while they drink, so puddles are perfect for this purpose. Even the smallest indentation in landscaping rocks or sidewalk pavers that collects falling rain can supply needed water; think of it as a teacup for tiny pollinators. Other times butterflies drink water from the surface of leaves. You'll see them lapping liquid from wet foliage with their proboscis. (See Plants That Capture Raindrops and Dewdrops, page 115).

Sometimes a moist human will do. Visit some of the hot, humid butterfly houses found at zoos and botanical gardens, and you might find yourself with a hitchhiking traveler or two landing on your arm or head. You are probably a very nice person, but they don't really like you as much as the salty sweat they intend to lick off your skin. Stand still and you can see the butterfly unfurl its proboscis and place it against your skin for a little taste.

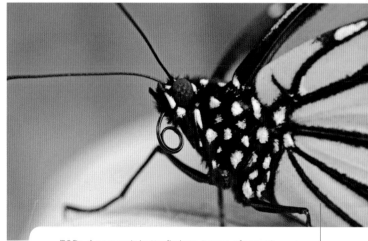

TOP: A monarch butterfly laps traces of recent rainfall from maple leaves.

BOTTOM: Monarch butterfly looking for a salty snack of minerals and sweat on this human hand.

There exist in some places around the world butterfly species that even feed off of tears, mostly those of animals. Just be glad you won't encounter a species like the madrilenial butterfly of Spain, sometimes known as the vampire butterfly, that likes to suck fresh blood.

Many butterfly species look for supplemental nutrition from carrion (animal carcasses) and dung. Butterflies in the carrion-eating food guild have a proboscis that's shaped differently for drawing the fluids from such substances. It may not look palatable but the butterflies are able to extract valuable nourishment from the decomposing flesh or fresh manure in the form of salts, carbohydrates, proteins, and amino acids. It's not unheard of to find large groups of butterflies, such as the gulf fritillary, feeding upon dead deer by the side of the road. At other times, you may find a butterfly dining on bird droppings.

FEEDING BUTTERFLIES

It may be hard to believe, but feeding birds is a billion dollar business. Why should birds get all the goodies when it's just as easy and fun to feed butterflies? Chances are you've already got some of their favorite foods on hand, such as those blackened bananas over on the kitchen counter.

Do:

- Use a shallow container for "fruit salad." Choose surplus and overripe fruit, such as oranges, bananas, and peaches.
- Put watermelon rinds to good use around the garden, placing them near forage areas. Whole melon slices of watermelon, cantaloupe, and honeydew will be appreciated even more.
- Freeze a banana, then thaw it out the next day. When it's gray and mealy, mush it up in a saucer with 2 tablespoons of blackstrap molasses and $1/4$ teaspoon dried yeast. Some folks add a little beer. Set out for butterflies, such as red admirals and painted ladies.
- For captive butterflies in classrooms or exhibitions, feed with Gatorade, Juicy Juice, honey water (one part honey mixed with nine parts water), or artificial nectar available from Monarch Watch (see Resources). Pour into a shallow container with a thin sponge or pot scrubber so butterflies can stay dry while feeding.
- Place feeders near windows and seating areas so you can observe and photograph the butterflies while they are busy feeding.

Don't:

- Use granulated sugar and water for "butterfly bait." Although many blogs and websites around the Internet suggest a sugar-water solution, it is not suitable for butterflies. It will gum up their proboscis.
- Use flowers from the florist. Many are bred specifically to contain little nectar. What nectar they may have won't remain viable after being cut.
- Forget to enjoy the elegant beauty of butterflies!

Monarchs in captivity feed from sponge soaked in sugar solution.

The painted lady, admiral, question mark, red-spotted purple, mourning cloak, and viceroy butterflies are fans of rotting fruit. They can be found feeding on fruit tossed into compost bins or on the ground beneath fruit trees whenever orchard windfall is plentiful. And just as there are reports of birds getting drunk from the fermented fruit they gobble from trees, so are there stories of tipsy butterflies feasting on fruit way past its freshness date.

The mourning cloak butterfly consumes tree sap as the bulk of its diet. Emerging from hibernation already as an adult in early spring, the mourning cloak is right there ready to find sap on the move oozing from trees as temperatures warm. The dark colored mantle allows them to peruse tree trunks undetected while looking for the sticky food. They walk down the trunk upside down while feeding on the sap that seeps from woodpecker holes and other wounds to the trunk. Other species, such as the red admiral and hackberry emperor, supplement their diet with tree sap as well.

Last on the list of funny food preferences is that of the harvester butterfly whose larvae are actually carnivorous: they crave aphids for dinner. The harvester butterfly hangs around the same location once it becomes an adult and eats the sugary secretions of the same little pests, tucking into a delicious meal of aphid honeydew.

PLANTS THAT CAPTURE RAINDROPS AND DEWDROPS

For pollinating insects, a tiny drop of water is all it takes for a drink. Once the sun comes out after a storm, you'll often find bees and butterflies sipping water from raindrops left on leaves.

Certain plants are better than others at capturing rainwater or gathering condensation. For example, *Alchemilla mollis*, or lady's mantle, is known best for the sparkling droplet jewels that catch on its scalloped

TOP: The dense hairs on leaves of lady's mantle, *Alchemilla mollis*, are especially effective at capturing raindrops.

MIDDLE: Raindrops bead up on the waxy surface of succulents.

BOTTOM: Glistening raindrops on the foliage and blooms of *Sedum* 'Xenox'.

foliage. Its botanical name refers to alchemy, the magical process attributed to the plant in medieval times when people believed the pearls of moisture were made of "celestial water." Nowadays, gardeners just appreciate this plant for the enchanting effect.

It's fun to discover which plants in the garden excel at capturing and holding drops of water, whether it's from rainfall or sprinkler spray. On early morning walks in the garden, when conditions are right, you'll find shimmering dewdrops as well. Plants with dense hairs on the leaves, such as lady's mantle, seem to do better at suspending the water drops. Plants with fuzzy leaves, such as lambs ears and silver mullein, sometimes hold the drops for a while until a slight breeze breaks the tension on the water's surface and the wooly leaves are left damp. In other cases, the waxy surface of some plants makes for a perfectly smooth skin that helps maintain the cohesion of the water molecules so they bead up just like water on a freshly waxed car. Succulents and tropical plants with waxy foliage are at opposite ends of the growing spectrum, yet are both good candidates for catching raindrops.

Some plants actually capture and funnel water with their leaves so that wildlife finds them useful as a drinking fountain. Bromeliads, the popular houseplants related to the pineapple, direct the water with curved leaves into a sort of holding tank for the plant. The little pool of water that collects at the center where the leaves join serves as a water trough for tiny creatures. Some *Colocasia*, or elephant ears, cultivars with cupped leaves collect water, then bend and spill it out ready for another free refill, of rain. Their names, 'Coffee Cups' and 'Teacup', say it all, but there's another smaller variety called 'Bikini Tini' that holds water just as well.

It's not only tropical plants that have this knack for holding water. Lots of hosta varieties can catch water among the dimples and folds of their foliage; however, *Hosta* 'Abiqua Drinking Gourd' leaves no doubt about its unique ability to hold water.

Native to prairies, *Silphium perfoliatum* or cup plant channels water into cups that are created where its leaf and stem meet up, offering birds, pollinators, and other wildlife a drink while feeding on it flowers or seeds. It should be noted that birds often hang around this watering hole waiting for these same thirsty critters and then call them dinner.

PLANTS THAT CAPTURE RAINDROPS

Venture out into your garden after the next rainstorm and see if you can identify more plants that successfully capture and hold raindrops.

NAME	SCIENTIFIC NAME
Bromeliad	*Bromelia*
Columbine	*Aquilegia canadensis*
Daphne	*Daphne odora*
Elephant ear	*Colocasia*
Lady's mantle	*Alchemilla*
Lotus	*Nelumbo nucifera*
Lupine	*Lupinus*
Nasturtium	*Tropaeolum*
Plantain lily	*Hosta*
Rodger's flower	*Rodgersia*
Sedum	*Sedum*
Spurge	*Euphorbia*
Smoketree	*Cotinus*

TOP: Small pools of water on lily pads offer pollinators a place to drink.

BOTTOM: Water droplets collect in the emerging foliage of *Hosta* 'Albiqua Drinking Gourd'.

GUTTATION: A DIFFERENT KIND OF DROP

Sometimes you'll see glistening beads of water decorating the edges of particular plants in your garden, with a drop of water balanced like an ornament on every leaf tip. The effect is fascinating but the name for the process is less than appealing. *Guttation* occurs when the soil is saturated and humidity is high. All of this moisture with nowhere to go creates root pressure in the plant, so the plant exudes the excess water through miniscule openings on the leaf margins called *hydathodes*. This happens at night so the resulting droplets are seen the following morning when bees are just beginning to make their rounds. The liquid seeping from the plants also contains xylem or plant sap along with minerals and salts. In fact, when these drops dry, sometimes you can see traces of salt on the leaves. Bees out foraging first thing in the morning are obviously on the lookout for pollen and nectar, but also for water for replenishing the supply needed to maintain the brood cells. So plant guttation is just one more water source that happens to be beautiful as well as useful for pollinators.

Guttation drops forming on the tips of fern foliage.

SAFE WATER

No matter where the water comes from, whether it's the sky or the sprinkler system, the most important question surrounds the safety of the water. Just as humans need water free from contaminants and dangerous chemicals, so do pollinators. You have control in your garden over the use of pesticides. However, bees encounter lots of water sources beyond your backyard as they go about their foraging duties. Pesticide drift into water sources can happen just as easily in your neighborhood as in areas near farm fields. New systemic pesticides now mean the poison is inside the plant, bringing about a whole other set of increasingly complex threats to pollinator health and wellbeing. Although the dose may be what is called sublethal, meaning it doesn't outright kill pollinators, it can cause disorientation that prevents bees from finding their hives and from communicating vital information about forage location or quality. More on the threat of pesticides and their usage can be found in Chapter 8.

Pollinators need safe water sources free from pesticides and contaminants.

"FEEDING" HUMMINGBIRDS

If you're slow to get the hummingbird feeder refilled, you might just catch an earful. It's not unusual for hummingbirds to fly up to the feeder and let out a string of short, sharp chirps that seem to scold you for your laziness. It's not surprising since the sky-high metabolism of these tiny birds means they live in an almost constant low-blood-sugar crisis. You'd be cranky, too. Sheri Williamson, co-director of the Southeastern Arizona Bird Observatory jokes that a hummingbird's vocabulary is 100 percent swear words.

But then almost everything about these little birds is extreme. They flap their wings around 80 times a second. They can fly as fast as 60 miles per hour. Their heart beats over 1,200 times a minute. All this action means they have to eat a lot and frequently. They may visit as many as 1,000 flowers per day consuming the sweet nectar that powers their bug-seeking activities.

Naturalists first discovered that hummingbirds in captivity would eat sugar water from containers as early as the late 1800s. By the early 1900s, various scientists and birding enthusiasts were experimenting with different vessels—glass bottles and test tubes fitted into a number of apparatuses. In 1950, the first commercially made hummingbird feeder was put on the market, and within a decade, the feeding of the little birds proved to be a new popular pastime.

Hummingbirds don't know instinctually to feed at the funnel-shaped ports, but they quickly learn to associate the feeder with the sweet syrup they crave. While flower nectar has about 12 percent sugar content, the sugar water solution in feeders comes in at 25 percent. And although they drink heavily from feeders, it remains only a supplement to their natural diet; they will still feed on flowers that provide the vitamins and minerals they need. They can be seen flitting back and forth between nearby flowers while visiting the feeder. And with each meal some of that food is stored in fat reserves to keep them from starving when it's cold or they need to rest. At the end of the day, the little bird that weighs only a tenth of an ounce will have consumed half its body weight in sugar.

THE RIGHT SOLUTION

Current thinking on health and wellness tells us to cut down if not avoid sugar altogether. Perhaps it's only human for us to want to assign those same values to the animals in our lives. But it's one thing to adjust your dog or cat's diet and totally another to try and make hummingbird food "healthier." Plain and simple, they need sugar and lots of it. When it comes to supplemental feeding for hummers, the motto should be "accept no substitutes." The time-tested recipe for hummingbird food is almost too simple, 1:4, one part sugar to four parts water.

Most people by now have been convinced to cut out the red dye once thought necessary to attract the hummers' attention. Still many folks like to see the feeders glowing with the red jewel-like liquid. Red dye No. 40 is a synthetic product that has no business in bird feeders (studies are inconclusive but common sense tells us to ditch the dye). Especially when you see how many hummers fight over feeders filled with crystal clear sugar water, why add one foreign ingredient?

Nectar concentrates offer an easy way to refill those feeders without all the gritty, sticky mess of making it from scratch, if you call mixing sugar and water "scratch." Make sure you are buying concentrate for convenience, though, and not for the promise of "high energy sucrose," which is, simply translated, sugar. Any other additives, such as protein, are unnecessary. Birds in captivity in a lab or rehab setting are the only ones that might need supplemental protein to make up for lack of naturally occurring insect meals.

FUN FACT

As they migrate northward in spring, hummingbirds drink tree sap from wounds in bark created by sapsuckers and woodpeckers.

DR. DONALD MITCHELL

Donald Mitchell has an M.S. degree in Conservation Biology from the University of Minnesota and has conducted field studies of hummingbirds and the plants they pollinate in Minnesota, Wisconsin, Colorado, and California. He is a federally permitted hummingbird bander and has served as vice president of the Minnesota Ornithologists' Union. He is a University of Minnesota Extension Master Gardener and attracts hundreds of hummingbirds annually to his garden near the Mississippi River in Red Wing, Minnesota.

Q. Why hummingbirds? What in particular about them fascinates you?

While I was a biology major at UC Berkley, I took a summer field course in the Sierra Nevada that required a pollination project. I chose a plant at random, a really tall delphinium that was pollinated by bumblebees and also the Calliope hummingbird. The Calliope male has an iridescent throat striped like a candy cane. I loved watching them visit that patch of flowers. These tiny birds mesmerized me and I was hooked. Large animals like lions and elephants that inspire conservation groups are called "charismatic megafauna." I like to say that hummingbirds are "charismatic microfauna."

Q. Are they as threatened in the same way as bees and butterflies? Are there other concerns?

There is no major threat to the group as a whole. There is no direct threat from insecticides, but there may well be an indirect threat since hummingbirds eat insects. However, climate change may be an issue for some birds. Changing climate has been implicated in an earlier bloom time of nectar plants at higher elevations in the western US that broad-tailed hummingbirds rely on when they return to their breeding grounds. So the bloom and arrival times are becoming increasingly out of sync.

The most dangerous time for a hummingbird after fledging is that first migration south. It doesn't follow a flock or have a parent to show the way. It just kind of points at a compass setting and starts flying. The lucky ones follow a direction that takes them by good food sources. If it does find a good route, once it's completed that cycle it can follow the same route again. Then there's a higher survival from year to year after that.

Q. Tell us about your involvement with banding hummingbirds, what do you learn?

The most important finding in banding studies is learning more about the demographics of bird populations and specifically when you band birds on their summer breeding grounds. They tend to display some amount of site fidelity, meaning they tend to return summer after summer to the same location. When I recapture birds that I've handled in previous years, it just astounds me that this tiny little organism has made it all the way down to Mexico on its own and back under its own power. With banding, we've also found out they display fidelity to migration routes and wintering areas.

Q. Plants or feeders? Does it matter?

I have feeders and flowers. Certainly, if I didn't have the feeders I wouldn't get the huge numbers of hummingbirds in my yard. However I would much rather use flowers simply because I enjoy the gardening aspect and I get a bigger kick out of watching hummingbirds visit flowers I planted for them and most likely chose and propagated myself.

Q. What tips do you have for feeders?

I mix up a large gallon jug of solution and use warm water to dissolve the sugar more easily. Then I store that in the fridge. I don't boil the water because once you put the solution out in the feeder the very first hummingbird that dips his bill in the feeder is going to inoculate the solution with bacteria. So it's an additional step that really isn't necessary. It's much more important to keep the feeders clean at all times. A lot of people get excited at the beginning of the season but get less attentive as time progresses. Keep the solution fresh and the feeders clean throughout the season.

Q. Do feeders discourage hummers from migrating at the right time?

No, because hummingbirds receive their cues to migrate not from dwindling food sources but rather the shortening of the day. So as soon as the days start getting short, that gives them the cue that it's time to start putting on fat to get ready to migrate. In early September I have dozens of hummingbirds crowded around my feeders tanking up for migration. At a certain point, they just leave on their own.

Q. What's in your yard?

Aquilegia canadensis, native columbine, is a good early season food source. *Lonicera sempervirens*, or coral honeysuckle, is also an early one. *Monarda didyma*, scarlet bee balm, is one of my favorites. *Silene regia*, scarlet catchfly, is a good one, and it's super easy to start from seed. *Lobelia cardinalis*, cardinal flower, is another favorite for a couple reasons. It can handle some shade and less than well-drained soil. Plus it blooms in late summer, which is the best time of the year to have hummingbird-attracting plants when you have a population explosion, all the fledglings and adults all fueling up for the southbound journey.

Regardless of all this, confirming the huge affection folks feel for hummingbirds, there are still people with purely good intentions wanting to make the nectar more wholesome. Some people boast they only use organic sugar, others of raw sugar. Hopefully people heed the plentiful warnings to never use honey, fruit gelatin, fruit juice, or brown sugar. And whatever you do, no artificial sweeteners! In the end, white granulated sugar that closely approximates the same chemical makeup of flower nectar seems to be preferred by hummers despite our well-meaning motives.

Although the jury is out, some hummingbird experts think distilled water might lack needed minerals and salts present in tap water. With that in mind, you should never add table salt to any hummingbird solution, even if you use distilled water.

It's simple and economical to make your own nectar. Arguments abound over the need to boil the water or not. Most contamination comes from the bird's bill rather than the water or sugar. If you're making large amounts of nectar, some experts think boiling is a better option. Mostly try to make no more than your hummingbird crew can use in a week, keeping any extra refrigerated. Whether in the fridge or the feeder, when the nectar gets cloudy, that means it has spoiled.

FAVORITE FEEDERS

There are two main types of feeders for hummingbirds, inverted bottle or basin, yet there are hundreds of variations on those two styles. What's the best style for hosting hummingbirds? The answer is the one that's easiest for you to disassemble, clean, and refill. By their very nature, hummingbird feeders are a sticky proposition and their maintenance can be bothersome—so be sure to find a feeder you can live with.

Glass feeders come in many forms: ornamental bottles, blown glass bubbles, potion-style pots, covered bowls, and more, in every color of the rainbow. These feeders are usually hung on poles or hooks around the garden. Some are attached to garden stakes you can situate in containers or throughout the flowerbeds. Red flowers in glass, metal, or plastic often form the feeding ports. These decorations suffice to attract the birds instead of dye. You can choose between feeders with a single port to one with multiple feeding opportunities depending upon how many birds visit your garden. Plastic feeders have the advantage of being less expensive, so it's possible to place more of them around the garden. They are lightweight, so they also usually have a larger capacity.

People are of two persuasions about perches on feeders. Hummingbirds are barely able to walk or hop on feet designed more for perching. After all, they spend 80 percent or the better part of the day perching. Providing perches saves them from expending energy hovering while feeding, effectively giving them a break. But some people find the perches block their field of vision and get in the way in photographs.

Feeders should be scrubbed out with hot water and a bottlebrush with every refill since they tend to accumulate black mold and other residues. Most experts discourage the use of soap that could leave residues. Instead they advise once every few weeks disinfecting with a bleach solution of $1/4$ cup bleach to one gallon of water. Feeders with only glass and metal parts can be run through the dishwasher on the sanitize setting. Feeders with plastic parts need to be washed by hand. The fewer parts to the feeder, the simpler the task.

LOCATING THE FEEDER

Place the feeder where you can best enjoy the antics and aerial maneuvers of your hummingbird guests while they fuel up. This may be outside the kitchen window or just off your front porch. If you're worried about glass strikes, apply stickers on the window or use blinds. Whenever possible, locate the feeder in partial shade since warm temperatures spoil the feeder contents faster.

The ideal feeder location has plenty of hummingbird-approved flower favorites planted close by as well a few places for perching. From there they can

TOP: A hummingbird perched between sips at an inverted bottle style feeder.

BOTTOM: Hummingbirds flit back and forth between flowers and feeder.

watch for tasty insects and unwanted interlopers. As the season begins, males are in competition for females and food, and once they claim their territory, they'll defend it vigorously. They can be quite aggressive in protecting the $1/4$-acre patch they are known to stake out. Females determine their territory once they have mated and built a nest. They don't want the flashy males around drawing attention to their vulnerable young, so they'll fight them away with beak and claw if needed.

Males use a number of strategies to ward off other males entering their territory. At first they may just fluff up their feathers and flash their iridescent gorget, the conspicuous red spot on their throat, in a show that communicates their robust health. If that fails, they may dive-bomb their opponent or give chase on wing, running them off to someone else's yard. At other times, it can get physical with actual fighting.

This behavior may seem like bullying but it's just the natural scheme of things in the hummingbird world. Yet if all this posturing ruins the peaceful scene you expected, you may have to take action. Provide more than one feeder and place them close together where they all focus on feeding rather than fighting. Or try locating the feeders in different locations for an out-of-sight, out-of-mind approach. If nothing helps, you may have to treat them like squabbling children and let them work it out on their own!

UNWANTED VISITORS

Hummingbirds aren't the only wildlife in the garden that appreciates a sure source of sugar. Ants, bees, and wasps may all show up and partake. If uninvited guests such as these are a problem, there are several things you can do. Change out bottle feeders for basin style feeders. Bottle feeders, no matter how tightly fastened will always leak and drip a bit in warm weather due to pressure changes. If ants are a persistent issue, purchase feeders with an ant moat since dead ants floating in the syrup put off hummers. If bees or wasps come looking for sweet stuff, you can try removing or painting over any yellow parts of the feeder that may attract them. Sometimes simply moving the feeder to a different location, even just a few feet, solves the problem. Don't use oils or sticky substances, such as petroleum jelly or duct tape, to trap pests as hummingbirds may brush against the material and get it in their feathers.

Cats, both domesticated and feral, are the most common threat to hummingbirds while they are visiting feeders. If cats are a problem in your garden, hang the feeder at least five feet high and in a place without cover for a stalking cat. Keep your own cats indoors if you value your hummingbirds' safety.

Note: Bats are perhaps not unwanted but definitely unexpected visitors to the hummingbird feeder. Luckily, bats happen to be pollinators, too! In areas of the Southwest, there are two species of nectar-eating bats that show up in springtime following the blooms of agave and saguaro plants. During this time, they often find hummingbird feeders irresistible. They can drain the feeders overnight and make a big mess of it, too. The bats don't hover like hummers but fly up, pause and quickly sip from the feeders with their incredibly long tongues before losing altitude. If you find the bats fascinating you could put out extra feeders or modify existing feeders to have bigger ports to fit their feeding method. One of the bats is on the endangered list, and the other is a species of concern, so they could use the help. If you only have eyes for the hummingbirds, bring in the feeders each evening. But be prepared to return them very early the next day when hummers are out foraging with a fierce hunger after sitting it out the night.

Sleeping bee attached to plant stem.

CHAPTER 7

SHELTER IN THE GARDEN

HOW POLLINATORS REST

If you've ever tried to take pictures of pollinators, you know it isn't easy. Pursuing them as they zoom or zigzag around the garden can be exhausting. Intent on feeding or finding a mate while escaping possible predators (such as the giant with the camera shadowing their every move), they hover or flutter for a moment and then they're on to the next flower. That term "busy as a bee" is every bit true and applies to almost all pollinators. Do they ever get a break?

For pollinators such as bees, butterflies, and even hummingbirds, rest and sleep are inextricably linked with the air temperature around them. As cold-blooded insects, bees and butterflies have a body temperature that is dependent upon their surrounding environment. And although hummingbirds are warm-blooded, being the smallest warm-blooded creature on Earth means maintaining enough body heat is an ever-present problem.

Warm-blooded animals generate internal heat to stay warm. They have to eat much more than cold-blooded animals to convert food into energy to maintain this rarely fluctuating constant temperature. Bees, butterflies, and other cold-blooded (or, more accurately, *ectothermic*) animals don't need to eat as much but rely on the environment and certain behaviors to regulate their body temperature.

This means these energetic pollinators are like tiny solar-powered machines buzzing and flittering around the garden. On warm, sunny days they are most active, able to go about their day sipping nectar and gathering pollen. But when it's dark or cold, they have to find a place to rest not only where they are sheltered from the elements but also safe from predators. To keep from freezing or starving while inactive, they are equipped with marvelous mechanisms that enable them to rest, sleep, and ultimately survive.

SLEEP

Honeybees have it pretty easy in that they don't have to look for shelter every evening when they retire from their day's work. They return to the hive when the sun goes down or temperatures drop below 50 degrees Fahrenheit. But since they are social insects, back at the hive there's a set role to perform even when they slow down for the evening. Talk about peer pressure, when you've been out foraging all day and just want to get off your pollen-packed extremities! Hive mentality means the duration and location of their sleep is determined by their age and job description.

Just hatched "cleaners" up to three days old sleep in the cells at the very center of the hive where's its warmest and there's less of a chance of being bumped and awakened. However, short bursts of sleep amounting to only about 30 seconds are what constitutes a nap. "Nurses," aged 4 to 12 days, don't have it much better since tending larvae is a never-ending task. They sleep a little longer, but don't always do so in the cells. Moving further away from the center of the hive, very few "storers" sleep in the cells, although they do get to sleep in longer increments. Finally, the "foragers" sleep on a more regular basis at the outer edges of the hive. Theories say they stay out of the main cluster for hygienic reasons, to keep diseases and parasites they may have picked up while outside away from the brood cells.

Individual honeybees may be cold-blooded, but the sum total of bees in a hive is described as a super-organism that some call warm-blooded since they use a collective shivering mechanism to regulate heat in the colony. The center of the hive where young are raised remains the warmest, with bees at the outer edges struggling to maintain an optimal temperature. Bees are also said to blur the boundaries and become temporarily warm-blooded because they can disengage their wing muscles (so they don't move) and shiver to warm up their thorax for flying without help from the sun. This same shivering can be used in reverse like a fan to cool the hive on a hot day.

With bumblebees and most solitary bees, sleeping accommodations are determined by gender. Females return to the nest at day's end to tend to their brood cells and rest. But perhaps you've found a bee nestled in a flower, remaining so still you think it's dead. Chances are it's only sleeping, and furthermore, it's highly likely a he. Males have no role in the nest, so they are left to do the bee version of couch surfing, finding a comfy bed wherever they can.

Tucked among the petals of flowers is just such a spot as it can be ten degrees higher than the outside temperature. They're fond of tubular or inverted bell-shaped blossoms such as morning glories or trumpet vine that cradle them as they sleep. Flowers with multiple ruffled petals sometimes don't hold much nectar but help to conceal them from predators while they doze. Otherwise they seek out sites under leaves or simply perch on stems. To keep from falling, bees can clamp their legs or mandibles onto the plant and somehow manage to stay attached even in their relaxed condition—often dangling precariously from a plant.

TOP: Honeybees tending brood cells: do they ever get a break?

BOTTOM: If you find a bee nestled inside of a flower, chances are he isn't dead, but merely sleeping.

To manage with loss of body heat while resting or sleeping, these bees enter a physiological state called *torpor*. Torpor is a reaction to stress or fatigue that produces lethargy. During torpor, body temperature drops and heart rate and metabolism slows. Torpor can last a few hours to a few days depending upon the situation. Like a mini-hibernation, this form of sleep helps animals to conserve their valuable resources. You might notice a groggy-looking bee early in the morning when you make your way around the garden first thing with a cup of coffee. While you wait for the caffeine to kick in, the bee is counting on the sun's heat to re-animate him. Check back a little later to see if he's gone or maybe you'll get to witness his waking up. Often when waking you'll see them do a kind of cleanup routine, wiping their antennae and legs in order to get ready for another busy day.

Butterflies face many of the same challenges as bees when it comes to catching z's. Dependent upon the sun's warmth to move, butterflies become less active when it's cold, cloudy, or dark. They are said to be quiescent when they are at rest, but not exactly asleep. When searching out a suitable rest stop, they search for a stable, sheltered location out of the wind. The resting spot needs to be hidden as well since they are especially vulnerable to predators in this sluggish condition.

They'll seek out a leaf or crevice among rocks, or maybe tall clumps of grasses where they will hang upside down to conserve energy. Some butterflies will choose a spot that receives early morning sun so they can wake and warm up earlier to forage before the competition.

RECHARGE

Early in the morning or on a chilly day, you may find a butterfly with its wings spread flat on a stepping-stone or flat rock, and nearly step upon them. Or perhaps you'll see a butterfly with its wings stretched out seemingly stuck to the south side of a fence or building.

During this activity, called *basking*, butterflies use their wings like solar panels to absorb the sun's rays so they can warm up to start their day. Butterflies fly best in temperatures between 75 to 90 degrees Fahrenheit, but to reach their flight threshold, a butterfly's wing muscles need a minimum temperature of 55 degrees Fahrenheit. This is why they'll often adopt this posture throughout the day, especially if it's on the cold side but still sunny. They're sort of recharging their batteries, so to speak. Most often, you'll find butterflies with darker colored wings such as

TOP: Pollinators can find hiding places within lush foliage of plants such as this allium.

BOTTOM: Rocky niches near forage add habitat value as well as charm to the landscape.

TOP: A red-spotted purple butterfly basks on gravel with wings spread to catch the sun's rays.

BOTTOM: As the sun sets, hummingbirds must find a protected place to sleep.

black swallowtails and mourning cloaks using this method of basking.

Other butterflies, such as sulphurs and satyrs, with light-colored wings but dark-colored undersides, use lateral basking to achieve the same result. They fold their wings together vertically and position themselves with wings facing the sun for maximum heat absorption. Light-colored butterflies use another take on this practice, using reflectance (reflected light) to warm their wings.

Just like its cold-blooded counterparts, the warm-blooded hummingbird also enters torpor in order to sleep without freezing or starving. Its heart rate slows to just a fifteenth of its normal rate. This is a deep sleep bordering on hypothermia that can easily make the hummingbird appear dead. (In fact, sometimes unhealthy or stressed hummingbirds don't survive this near-death dance.) Before they sleep, they'll hunch down into their feathers on a favorite perch in a tree or shrub. The female may sit on the nest with her young while sleeping. But watch out, although it may take the tiny hummingbird as long as 20 minutes to emerge from this heavy slumber, it does so with a raging hunger, consuming 25 percent of its daily food requirement immediately afterward.

HIBERNATION

Short states of torpor satisfy daily rest and sleep needs for pollinators during the growing season. When colder weather approaches, pollinators have two choices. Just like humans crave comfort food and down comforters, or better yet a tropical vacation, pollinators increase food intake and then hibernate or leave town for warmer locations as winter arrives.

Honeybees work all season long in the hope they've made enough honey (anywhere from 30 to 80 pounds depending upon geographic location) and stockpiled enough pollen to make it through winter. When temperatures drop to around 50 degrees Fahrenheit, honeybees stick close to the hive and prepare to huddle. They produce a resin-like material called *propolis* to seal any cracks and weatherproof the hive. Next they form a vibrating ball of bees called a *winter cluster*. The queen sits safe and toasty at the very center. Bees close to the center shiver simultaneously to warm the hive up to around 65 degrees at the center. Bees on the outside of the cluster stay still to act as an insulating layer. They regularly exchange places to warm themselves and feed upon the stored honey. If the bees are stressed by extreme weather, predator invasions, or other disturbances, they'll react by eating more, throwing the hive out of balance. Beekeepers often help their hives along with sugar-water solutions and pollen cakes to ensure winter survival.

Other bees are on their own. Although bumblebees live as social insects, the colony simply dies out when winter nears, leaving only mated queens to hibernate through the winter. The queens remain in a lengthier state of torpor called *diapause* until they are ready to create the next generation once spring arrives. Many solitary bees as well as butterflies also enter diapause to sit out winter. Unlike torpor, insects can spend this form of suspended animation in various stages of metamorphosis either as an egg, larva, pupa, or adult, depending upon the species.

HOW POLLINATORS MIGRATE

As the season draws to a close, most bees and butterflies will hibernate in one form or another. Some will spend the winter tucked snugly in brood cells as an egg or pupa. Adults will find a warm and sheltered spot between rocks, behind loosened bark, or bunk down in a recycled rodent nest until spring returns. However, a few well-known and beloved species, such as the monarch butterfly and the ruby-throated hummingbird, head south in search of better weather and abundant food.

Some butterfly species, such as the painted lady, common buckeye, clouded sulphur, and question mark, will make a shorter one-way trip looking for new food supplies, but it isn't really considered a true migration. Rather, these forays are called emigration and are more a function of range expansion and contraction that follows availability of food.

MONARCH MIGRATION

Only the majestic monarch butterfly makes the incredible 3,000-mile round-trip odyssey between its summer breeding grounds and winter roosting sites. Completing this arduous journey is further toughened by almost insurmountable problems of habitat loss at both ends of and along the route, coupled with extreme weather. When folks speak of the miraculous monarch migration, it does truly seem a miracle.

The monarch migration has long been surrounded with mystery and still poses many unanswered questions. No one knows exactly how long the migration has occurred, but the general consensus is that it has been going on for thousands of years.

Until quite recently, people were in the dark as to exactly where they went. A portion of monarchs that

During migration, monarchs gather together in large clusters to roost.

DR. ORLEY "CHIP" TAYLOR

Dr. Chip Taylor is the founder and director of Monarch Watch, an outreach program focused on education, research, and conservation of monarch butterflies. The program engages citizen scientists in large-scale research in relation to the monarchs and has produced new discoveries about the dynamics of monarch migration. He is trained as an insect ecologist and is a professor in the Department of Ecology and Evolutionary Biology at the University of Kansas in Lawrence.

Q. Have you always studied monarchs?

I raised monarchs as a kid. Subsequently, I developed into a biologist, did work on butterflies early in my career, and shifted to bees, honeybees, and so-called "killer bees" for 22 years. I shifted to monarch butterflies in 1992.

Q. Why did you choose them?

Monarch butterflies are really great for educational purposes. I was teaching a graduate course and looking for organisms I could work into the course. As I got into it, I realized there was relatively little known about migration per se. This led to a tagging program, development of Monarch Watch, and all that we do. That ultimately morphed into a conservation effort.

Q. What's the purpose of tagging monarchs?

The original purpose was to determine where the monarchs come from that arrive in Mexico each year, but we wanted to get other data on the size and scope of migration, the success of the migrants originating from different parts of the country, and what the probability is of making it to Mexico. It's not equal for all the monarchs. The monarchs that originate from the Midwest are the most likely ones to get to Mexico. Those along the East coast, they'll take an offshore wind in the morning and get out to the ocean and if they don't get onshore in the afternoon, they may be stuck out there.

Q. Why is Texas so important to monarchs?

Texas is really the most important area because monarchs are coming north into Texas to breed, to produce the first generation. If you go further east there just isn't enough milkweed. What happens in the spring in Texas determines what happens the rest of the year for the rest of the country.

Q. What are your thoughts on the "*Curassavica* question," that is, an abundance of tropical milkweed possibly throwing off migration?

If you look at the larger picture, this is a very small issue. The larger issue is the loss of habitat. If we do not deal with an annual loss of habitat on the scale of a million acres a year, the population will continue to go down and it won't matter how much *Curassavica* we plant.

Q. Do you think gardeners can make a real impact for monarch conservation?

Yes, absolutely. Here's why. We have big holes in our habitats for monarchs. We're down to probably less than one or two percent of the total native habitat. This is why we have to counter large-scale fragmentation with small-scale positive fragmentation so that we have milkweeds everywhere we can possibly put them to ease the burden for butterflies in finding places to lay eggs. What do we do about that? We create Monarch Waystations. We create them everywhere we possibly can. We've got something like 9,000 waystations registered and we need something like nine million. We need gardeners all over the country to be involved here.

Q. Is there any good news when it comes to monarchs?

In 2014, we issued more tagging kits, had more butterflies tagged, the taggers had better experiences, we had fewer complaints of finding no butterflies, and so on. There's a good chance the population will actually be twice what it was last year but that's still a very small number. If you take a small amount and double it, you still have a small amount. That's what we're dealing with.

Q. What's in your yard?

I have an experimental plot in an old pasture where I'm trying to restore milkweeds. I'm trying to get as many species as I can in this plot that are native to this area. I had over 100 plants marked with flags, but I'm a little frustrated because grasshoppers ate the flags!

Creating monarch waystations in your garden is a resourceful and responsible small-scale solution to a large-scale problem.

summered west of the Rocky Mountains were discovered to spend winters in groves of eucalyptus, cypress, and Monterey pine along the cool, foggy southern California coast. But not until 1975 did researchers find out what villagers in a small area of Mexico had known all along. The largest population of monarchs, the ones that fluttered and fed throughout the Midwest and eastern part of North America, all headed to a specific geographical area to roost in the oyamel fir forests in the mountains of central Mexico.

According to their biology, monarchs need to roost in a cool place where they can conserve energy while protected from winter wind and snow. But this overwintering site needs to be even more specialized than that, a perfect combination of trees, vegetation, fog, and water sources. It turns out this relatively small area in the Sierra Madres on steep southwest-facing slopes, two miles above sea level, fits all the requirements. As deforestation has happened, these overwintering grounds have shrunken to even smaller percentages, making it hard to support large populations of monarchs at this end of the route. The last thing you need is a shortage of hotels after making that long of a trip.

To describe the endless loop of migration, you have to start somewhere. Jump in at early summer when you start to see monarchs fluttering around your gardens. These monarchs live for three to five weeks during which they visit flowers for nectar and seek out milkweed plants upon which to lay their eggs. The resulting caterpillars go through metamorphosis and emerge from their chrysalises ready to repeat the process. They immediately reproduce, yielding three to four generations throughout the summer. Then something different happens with the last generation.

The last generation of monarchs is not only physically different in their nervous systems and hormones but behaviorally different as well. There are various environmental triggers that make for this phenomenon. Scientists believe that a number of cues may account for the change; cooler temperatures at night, decline in host plant quantity and quality, but most essentially decreasing day length. They go into reproductive diapause, meaning they will cease to mate or lay eggs until the following spring. This generation is more focused than ever upon feeding, so intent you can approach them at a bloom and they don't even flinch much less flutter away. The accumulated fat they'll store in their abdomen in preparation for the journey must fuel the entire flight as well as keep them into spring. This is the most critical time to have an abundance of late-blooming flowers, annuals and perennials, native and introduced, in your garden to provide the nectar that will power these butterflies back to their winter homes.

Decreasing daylight tells them it's time to leave, but where's the map that tells them where to go? Once again it is thought to be a combination of factors leading them to their destination: the sun and stars, geographic landmarks, and possibly infrared energy. This internal GPS takes them back to the same place time and time again.

The migration starts in the north. Butterflies in the northwest head to overwintering grounds in California, while the larger numbers living east of the Rockies, in the Midwest, and in the eastern part of the continent head south and eventually funnel back toward Mexico. As they travel southward, the monarchs will stop at dusk along the way. Although they are not social insects, during the trip they cluster together in trees at night before starting out solo the next day, perhaps trusting that there's safety in numbers. They are the sturdier of many butterfly species, but they'll still fly an obstacle course through bad weather and predatory birds. To save energy, they coast on air currents and ride thermals until finally arriving at their winter roosting grounds, often to the exact same trees, probably tattered and definitely lighter.

Upon arrival at the overwintering grounds, the forest serves as a buffer to harsh weather conditions where the monarchs can be seen clustering on the branches and trunks of trees, even clinging to pine cones and needles, seeming to upholster the trees with their striking orange and black pattern. Temperatures can hover just above freezing, so the butterflies are in a state of torpor. Occasionally, they warm up and break free from the group to find a drink of water, but not for long. They remain there for a few months and then as days start to lengthen again, they grow active. Around mid-March, they start to fly north with one thing in mind: milkweed (for more on milkweed, see page 83). They seek out any and all types of flowers for nectar, actually gaining weight on the way, but the nectar only serves to fuel their flight in search of the one plant that can guarantee the survival of their young.

Finding milkweed stimulates the development of their reproductive organs, and they begin to mate and lay eggs on the plants. They'll usually lay only one egg per plant to ensure each new larva has plenty to eat, so obviously they need lots and lots of milkweed. Coming full circle the life cycle starts all over again. It's always cause for celebration when the first monarchs appear in your garden. Will you have the nectar plants and milkweed they need?

HUMMINGBIRD MIGRATION

Lots of species of hummingbirds migrate, but the ruby-throated hummer is the one you'll most likely see in your garden. It shares a similar migrating dynamic with many of the others. Although the Anna's hummingbird often wanders back and forth following food sources, it doesn't really migrate.

Some birds simply become vegetarians in the winter, but hummingbirds, as carnivores, have to go where the food is—and that's back to the tropics. Some may think they only load up on sugary nectar, but that's just to fuel their insect-seeking activity. The carb loading keeps up their high metabolic rate so they can catch flies, gnats, mosquitoes, aphids, and other insects. But they especially love to fatten up on spiders. Surprisingly, this favorite food makes up 60 to 80 percent of their diet.

Their instinct to migrate is triggered by decreasing day length rather than from cooling temperatures or lack of food. They bulk up before leaving, but people needn't worry that leaving out feeders will stop them from migrating. They actually leave while food supplies are plentiful, sometimes as early as mid-July. The migration peaks around mid-August to early September, though.

Hummingbirds are solitary birds, so they don't travel in flocks that might influence their final destination. A few stay along the Gulf Coast and the Outer Banks, but most will head toward southern Mexico and anywhere as far as Northern Panama. To get there, they have to cross the open waters of the Gulf of Mexico, leaving at dusk for a nonstop flight of 500 miles. Depending upon the weather, the trip can take 18 to 22 hours, an amazing marathon for such a miniscule bird, even though some theories say they stop to rest on oil rigs and fishing boats. And it's thought some just island hop all the way through the Caribbean. No matter how they make it, they'll arrive at only a third of their previous weight.

The return migration starts as early as January and is spread out over three months, with males leaving first. They travel an average of 20 miles a day, eating enough so that they actually gain weight on this leg of the journey. Older birds that have made the trip before are known to retrace their routes and return to where they hatched, often to the same gardens and more incredibly to the same feeders. That is a huge reason to plant more hummingbird favorites and put out feeders so as to build up your own backyard hummingbird population!

PROVIDING SHELTER FOR POLLINATORS

Picture yourself really small. Now picture yourself even smaller. To get an idea of what it's like for pollinators to go about their daily lives in your garden and in your neighborhood, you need to imagine yourself as a tiny bug. We humans can (hopefully) go inside when we're cold, sleepy, or frightened. However, pollinators look to their environment for shelter from the elements and potential predators. Even honeybees still have to maneuver through the hazards of the outside world before returning to the safety and warmth of the hive.

If you never thought of this Lilliputian aspect of insect life before, the mental exercise is a real eye-opener. Your yard is like a far-reaching territory that pollinators must travel around as they go about finding food, mates, and nesting and resting sites. And the typical home landscape is far removed from what they require to remain safe and protected as they do it. Sure, there's the circle of life: it's expected a certain number of pollinators will meet their fate of being eaten by birds or taken down by harsh

FUN FACT

Rumor has it hummingbirds hitch rides on the backs of other migrating birds. Not true, but they do fly with other migrating birds along their route.

A rose offers this bee a warm and sheltered place to rest within its petals.

weather. Yet conventional gardening standards add to those problems, making it much harder for them to overcome those challenges and maintain their numbers.

The flowerbed or shrub border where they forage and feed is just one of a series of islands dotted here and there, surrounded by an immense landscape covered with turf and hard surfaces between homes and other buildings. The lawn is an open savannah fraught with danger. That generously proportioned concrete driveway is a vast and barren desert exposed to all manner of peril.

It's difficult for pollinators to move safely through gardens with few plantings and little diversity of plantings within that. As discussed in earlier chapters, the commonplace landscape plan—lawn, three foundation shrubs, and a tree—repeated over and over, particularly in sprawling suburbs, offers pollinators little in the way of food. It's just as paltry a choice for cover from natural threats. The large, open spaces in between minimal greenery and acres of asphalt leave pollinators especially vulnerable as they navigate their limited world. Chances are they'll seek out another area with more advantageous vegetation and hardscape. But where?

Pollinators do use torpor as a survival strategy when sleeping, but they still need to find places to rest and shelter safely during the day, especially in cold or wet periods. Sitting out bad weather means they aren't eating, so they need to conserve energy and body heat while they wait. Sometimes they are caught out when sudden shifts in weather leave them unprotected; without immediate cover they can be tossed about by high wind gusts or drenched in downpours. This all acts to deplete their energy reserves and put them at risk.

Pollinators use a wide variety of defense mechanisms to avoid predators. Lots of bees flaunt their yellow and black stripes as a first warning and back it up with a powerful buzz. Other predators are spooked away by their stingers. This must be somewhat effective since minor pollinators, such as flies and wasps, have evolved to look like bees and borrow off this scary appearance scheme for survival. But it doesn't always work, there are plenty of predators that don't buy into it and hunt them anyway. With bees, most predators hang around the nest looking for a quick and easy bite, but there are other times as bees are out foraging when they also need to worry. Robins as well as some other birds are known to catch bees and simply rub their stingers off on a branch before snacking on the tasty insect. After all, there's a sweet reward with the honey-like contents in their second stomach, also known as a crop.

Bees are most vulnerable when they have used up their limited fuel supply while out foraging, particularly on chilly days. The bee must stop to recharge. You may see a bee sitting inside a flower or on a leaf wiggling his butt in an up and down motion. For this process, the bee has to stop and pump air through its abdomen in order to use the oxygen to warm up its wing muscles to fly again. They need a safe place to perform this behavior.

Butterflies might seem like a funny, fluttery thing to eat, but they have their fans. Besides birds, snakes, toads, frogs, lizards, dragonflies, and other animals are known to eat them in adult form or in various larval stages. The

Even the smallest crevice is a potential hiding spot or temporary shelter from bad weather.

insects use their looks even more than bees to scare off potential predators. The bright orange monarchs signal the toxic taste they acquire from milkweed to become unpalatable to birds. Other species employ enormous eyespots or other means of trickery and camouflage to avoid being eaten. In both larval and adult life stages, some species mimic twigs and other natural objects to blend in to the landscape. At other times, they gather together and hope they're in the majority that survives an attack.

You'd think hummingbirds move so fast nothing could catch them. However, domesticated and feral cats are their number one enemy. Hawks and other birds pose a threat as well to the tiny birds. Crows and blue jays sometimes raid their nests, devouring eggs and young.

Despite all the tactics and strategies pollinators use to avoid predators, often there is no alternative but to physically hide, and that demands a hiding place.

Some of the same materials and practices that provide opportunities for pollinator nesting sites also serve to offer shelter from weather and refuge from predators. And once again, many of the methods used to protect pollinators also have the added bonus of enhancing habitat for beneficial insects, birds, and other wildlife. It's hard to help one species without helping others.

Naturally occurring yard debris is valuable, so think twice before hauling it away. Leaving deadfall from trees in the form of snags, logs, stumps, downed branches, and brush piles offer an abundance of hiding spots. Leaf litter should be removed from vegetable gardens and other edible gardens to avoid carrying over fungus and disease pathogens from year to year. And it's necessary to rake leaves from your grass to maintain a healthy lawn, but fallen leaves directed into ornamental beds and borders instead of yard waste bags make great mulch that eventually decomposes and adds nutrients to your soil. The process goes even quicker if the leaves are shredded first and then reapplied. Just think of how many tiny creatures can travel undetected beneath a blanket of leaf litter.

PLANTINGS THAT PROTECT

The plants in your yard play a major role in creating cover. Just as landscapes with optimal forage require greater plant diversity and density, the same layers of landscape that offer more prospects for food do the same with opportunities for cover. Pollinators and other garden wildlife use the different layers of deciduous and evergreen plants at different times for their changing needs. In a successfully layered landscape, a broader palette of plant shapes and sizes offers multistory cover that pollinators can use like a staircase to move about the different heights of habitat. Bees use the garden top to bottom, foraging high up on flowering trees or cruising a carpet of dandelions for food, maybe nesting in woody stems among shrubs

This small front yard has great structural diversity and supplies multiple levels of shelter.

or under tufts of grass at or below the ground. Butterflies forage on flowers at eye level while their caterpillars might occupy the leafy tree canopy or crawl among violets that serve as larval hosts.

Within the layers, it's important to establish denser plantings. Permission to grow more plants? What gardener doesn't want to hear that? "Dense" doesn't connote crowded, but definitely fuller. Ignore the spacing rules on plant tags. Nature doesn't plant on eight-inch centers; it sows in clumps and drifts. This doesn't necessarily mean having to buy more plants either. Lots of perennials, ornamental grasses, and ground covers can be divided and replanted to increase the quantity of plants within an area over time. Utilize reseeding varieties for more free plants with each season. Leave them in place or relocate them; it's easy to move the small seedlings around the garden to where they are best suited. When plants are placed closely, they touch and weave together, creating a covered bridge effect that allows for safer passage on the ground as well as through the upper parts of the plant.

Evergreens and ornamental grasses by this water garden offer shelter while pollinators seek a drink.

The densest of plant configurations is a hedge. Whether it's a monoculture or mixed planting, this living wall is a fantastic device to define boundaries, screen properties for privacy, or simply make a graphic statement. Hedges can be deciduous or evergreen, or perhaps both. Best of all, thick, healthy hedges are an important source of cover where bees, butterflies, hummingbirds, and other small creatures hide out or huddle from the wind and cold. Hedgerows are the wilder version of the clipped hedge. Often used for enclosing livestock in the past, the old-fashioned fence was a mixture of bushes, brambles, and vines that grew knitted together forming an impenetrable barrier. Lush with flowers and fruits and full of cozy crevices, maybe it's time to bring back the hedgerow. How fun would it be to create a newfangled, tangled-up hedgerow and see how many pollinators flocked to its shelter and bounty?

It only makes sense to place cover plants next to food plants for when pollinators need to make a quick escape. The same goes for ponds, water features, fountains, bird baths, bee waterers, and other sources of moisture—animals don't feel comfortable drinking in an exposed setting. In naturalistic water gardens, tall plants, such as cattails, iris, and papyrus, and clumping plants, such as sedges, soften edges from a design point of view and also give pollinators a fast getaway if needed. For more formal water features, groupings of potted plants offer the same purpose in an attractive manner.

So many garden designs for pollinator gardens focus on pretty flowers without regard to other crucial needs. Butterfly garden plans—although as of late some are starting to take larval host plants into more account—are notorious for showing isolated bubbles of beautiful flowers while disregarding shelter requirements. To create more continuous cover, your beds and borders need to flow into one another or at least be only separated by short distances. This is a great time to consider reducing the area devoted to lawn and simply connect the dots so to speak with those planted patches.

There's also no law (only unimaginative habit) that says shrubs have to hug the foundation. Plant-rich borders abundant with small trees, shrubs, perennials, ornamental grasses, and groundcovers can line the edges of a lot or completely surround your property. Vary the depth of the borders with curving edges to let it meander naturally throughout the garden. Let it jump across pathways if possible, but leave some lawn in the middle to anchor the design and show observers or nervous neighbors the landscape is intentional.

Finally, hook up with your neighbor. You could install plantings on your side of the property line adjacent to next-door plantings that would appear to merge with them, and effectively enlarge and expand the habitat area without actually encroaching upon their space. This could be done with or without the neighbor's knowledge or cooperation, but wouldn't it be lovely to conspire with them to add more habitat in the same way? On larger properties abutting natural areas, blur the lines and have native plantings blend into the scenery around the edges. This would not only link the property with wilder habitat and provide for food, nesting sites, and cover but give the home an impression of being indigenous to the land and settled into its site.

The end goal with these denser plantings is to then shape areas of continuous cover that shield pollinators as they commute between forage areas and resting or nesting sites. This can be thought of as a just a smaller version of the proposed habitat corridors that thousands and thousands of pollinator-friendly gardens linked together can hopefully achieve throughout the country. Rather than making pollinators run (crawl, creep, flutter, of fly) a gauntlet across the wide open spaces of your yard, you want to connect plantings to build a better vegetation highway with fewer "miles" between eating places, hotels, child care, and rest stops, all within your garden. And once that's done, you might encourage the fellow next door to do the same and join the two habitats. What if an entire neighborhood followed suit and established a network of habitats? You can see where this is leading.

HARDSCAPE THAT HELPS

You may never have considered how the hardscape around your home helps shelter and support pollinators. But the more you learn to think in three dimensions, like a pollinator, suddenly you see how they use native materials and introduced structures and surfaces to their advantage. Hardscape takes many forms, both utilitarian and decorative, with the two purposes many times overlapping: buildings, fences, walls, driveways, pavers, patios, decks, pergolas, plant supports, sidewalks, steps and pathways, boulders and rocks, stepping stones and more. And just as a landscape with greater biodiversity enhances wildlife habitat, one with more structural complexity provides pollinators with more places to rest, warm up, or hide.

Nooks and crannies are probably the most appreciated features of any hardscape. Pollinators exploit the natural and manmade crevices in stone structures, rockwork, concrete, and brick, using them for nests, hideouts, and shelter from the ravages of weather. Dry-stack stone walls, laid without mortar, are especially good for this purpose: insects can burrow deeper into the structure, possibly using it for a hibernation site. Stone and brick also have the benefit of heat absorption that maintains a more even temperature inside the exterior or freestanding wall. Chinks and holes in wooden structures and fences afford similar spots to hide and nest but without moderating temperatures the same way.

It's not all about hiding though. Many pollinators look to fence posts, trellises, and other structural high points around the garden for lookout spots. You may think hummingbirds remain in perpetual motion, but surprisingly they spend a good deal of their lives perching. Hummingbirds hunt

insects for protein from perching spots. They will also seek a perch that serves as a watchtower to spot danger or see rivals approaching. Being very territorial, some will set up shop on a perch to defend a particular feeder. But mainly they like to use the perch as a home base to rest in between feedings and hunting expeditions. It's not unusual to see hummingbirds sitting atop the tall stakes used to prop up the same dahlias they like to sip from late in summer.

Rather than marking territory, the males of such butterfly species as mourning cloaks, black swallowtails, and red admirals use perches as a vantage point to check out possible mates. Simply inserting stakes, branches, or bamboo poles into the ground throughout butterfly forage areas provides perching spots for them as well as more opportunities for you to observe butterflies sitting still for a moment!

Another way in which you'll see pollinators utilizing hardscape in full view is when they use sun-warmed surfaces to recharge. Butterflies take advantage of east- and south-facing walls of buildings for basking. Other times you may find them with wings spread flat upon rocks, stepping stones, or pavement soaking up rays. Different species may prefer dark surfaces that absorb heat while others like light colored surfaces that reflect warmth. Be careful not to step on these basking butterflies. Place some flat pavers and stones in sunny areas among the garden where foot traffic won't be a threat.

Rock gardens with a wide variety of season-long bloom are a natural for great pollinator habitat, with different configurations of rocks favoring various species' preferences for nesting and hiding places. Some will hide underneath rocks seeking damper conditions, while others will go for snug gaps between rocks. Adding rocks, boulders, and flat stones to any existing perennial borders and beds is a good step toward enhancing habitat value.

When thinking about hard surfaces such as patios or paving in the garden, permeable surfaces that allow water to percolate in place are a better choice. Rather than sending valuable water into storm drains where silt and debris enter rivers and streams, they act to absorb rain and storm water, filtering it into the ground right where it falls. From a design standpoint, they offer a creative alternative to large expanses of boring concrete or hot asphalt. They also break up long stretches of barren habitat for pollinators. For patios and pathways, avoid the use of solid mortar and consider porous

TOP: There are plenty of hiding places between the stones of this dry-stack wall in addition to the plantings.

BOTTOM: A hummingbird perches on a stake between sipping on dahlias

materials, such as sand and gravel, between pavers. Low, creeping groundcovers that tolerate varying degrees of foot traffic can fill crevices around stepping stones while adding charm and a sense of maturity to the landscape, especially if they are flowering varieties, such as creeping thyme, that bees adore. It may seem negligible, but even the tiny cracks between stepping stones are potential habitat.

MANMADE HABITAT

Sometimes nature needs a little help. Sometimes it's the other way around. While the practice of beekeeping for honey has been going on for thousands of years, people are discovering ways they can assist other pollinators, such as wild bees, butterflies, and beneficial insects, by providing manmade structures for nesting and cover. Backyard beekeeping is a growing interest among folks engaged in the local food movement and homesteading. But if you're not ready for that commitment, there are still plenty of things you do to help the cause.

BEE HIVES

It's not hard for anyone with a sweet tooth to imagine out why early man mastered beekeeping. But what clever guy figured out that smoke made bees docile, so much so you could harvest their honey without pain and suffering? Talk about a eureka moment! Since then honeybees have been kept in all sorts of hives. The first ones were simple mud and straw contraptions. After that, various styles of hives made of hollow logs, clay vessels, and baskets followed with different cultures making improvements upon the basic design as time went on.

Nowadays, two versions of beehives are most widely used in North America and many parts of the world: the Langstroth and the top-bar hive. It takes many books and classes to educate people on the practices of beekeeping, but for this book's purpose a basic explanation will do.

Back in 1852, a Reverend Langstroth from Philadelphia perfected the system of stacked open boxes that beekeepers used to house bees and harvest honey. At some point, frames had been introduced for the honeybees to build their comb upon, but since the bees would create sticky hard bridges between the undulating combs, whenever the beekeeper removed the frames, it would break and damage the entire hive. Langstroth discovered through trial and error the exact spacing of the frames that discouraged the bees from doing this—it's called the *bee space principle*. Once hives were built using this specification of $1/4$ to $3/8$ inches between frames, the beekeepers' life got a lot easier.

Allowing for subtle variations, a Langstroth hive starts at the ground with the hive stand, on which sits a bottom board. Some include a landing board. The first box is the brood box where the queen lays eggs, and directly above is the queen excluder which keeps the brood separate from the honey. The top box, where the honey is stored, is called the super. Worker bees travel back and forth between the two areas in their daily duties. An inner cover sits on top of the super and then an outer cover forms the roof of the hive. The beekeeper tends to the bees by lifting the covers and boxes enabling him or her to inspect the frames without harming the bees' handiwork. At the end of the

A beekeeper removing and inspecting frames on a traditional Langstroth-style hive.

season, a portion of honey is left for the bees as a supply for winter survival; the amount depends upon where you live and the severity of winters. Any extra is removed and processed with a centrifuge and then bottled.

Top-bar hives are thought of as a more natural way to keep bees. But if you asked the bees, they probably don't think there's anything natural about "harvesting" their honey. In days of yore, it was called "robbing" honey, in fact. The top-bar hive is a long enclosed box about the same size and shape as a baby cradle. The top lifts up on a hinge to expose a simple arrangement of bars, or follower boards, sitting across the top of the hive. The bees fasten their combs onto a small length of waxy starter substance and form natural combs from the slats. This type of hive is touted as easier for some since there's less heavy lifting and smaller amounts of honey to deal with. However, it needs to be attended to more frequently. To harvest, the beekeeper removes the whole comb and crushes it to release the honey. This hive is considered more hygienic since frames and combs that could harbor pathogens are not reused and it starts with a clean slate each season. Each hive has its merits, so potential beekeepers should do their homework to see which style fits their lifestyle and motivations.

MASON BEE HOUSES

Providing mason bee housing with nesting materials is a great way to raise bees without having to don that big white suit and veil. You won't get any honey, but your garden will benefit, and you'll have the satisfaction of helping out one of nature's most helpful creatures. Mason bee houses come in many configurations with almost as many adorable names to describe them.

Whether you decide to put up a bee abode, bee lodge, or an alpine-inspired bee chalet, there's no argument that mason bees are most welcome to set up residence in your garden. Mason bees are among the super pollinators of the bee world. They won't damage your home with their nesting activities, either. And their gentle nature means they rarely sting.

If you install one of these housing options in your garden, you can wait for existing bees to find the nests or you can help the odds by purchasing cocoons from a growing number of mason bee businesses (see Resources, page 170). Make sure that you purchase mason bees appropriate for your region, *Osmia lignaria lignaria* for east of the Rockies and *Osmia lignaria propinqua* for west. Ask the vendor to be certain.

These ambitious bees got their name from the method and materials they use to seal their tube-shaped nests. After forming the series of cells, laying eggs, and provisioning the individual compartments with food, they cap off the tube with mud. They go about this just like a mason uses mortar to join bricks and stone.

You would think mud is mud, but mason bees require mud of clay-like consistency. If your garden has sandy or loamy soil, mix equal parts powdered clay with your garden soil and water to form a paste. Dig a shallow hole near the nesting area and pack this material on the south side of the hole to keep it moist. Keep it damp during the nesting season.

The Right Bee Abode

No matter the bee house style you choose, they all offer the tube-like cavities Mason bees call home. Solid wood bee blocks are easy and inexpensive to make or you can purchase any number of prefabricated bee houses. With either kind you help the mason bees with a head start; the female bee won't have to search far for a prime nesting site and can get down to the business of laying eggs right away. After all, she only lives for about six weeks during which time she will lay 15 to 20 eggs.

Wooden bee blocks can be made with scrap lumber. Take an untreated section of a 4x6 board about a foot long. On the short side, drill rows of evenly spaced holes, 3 to 5 inches deep, taking care not to go all the way through. You can use different size holes, but $^5/_{16}$-inch works best for mason bees. Top the block with a roof shingle or small piece of flat board at a slope to help shed rain and keep it dry.

Wooden bee blocks do come with a caveat. After even one season, parasites and mold can build up, creating unhealthy conditions for the next season's guests. At the least, you'll need to clean out the block after the nesting cycle is complete. Redrill the holes to remove debris, then dip the block in a bleach solution and let dry to disinfect and make it ready for the next generation. Better yet, since they are cheap and easy, it's recommended you discard that block and make a new one to be on the safe side.

Paper straws or reeds bundled into coffee cans, clay tiles, or flowerpots are another inexpensive way to help mason bees. Avoid plastic straws since they are too slippery for the cell material to grip. These materials have the advantage that they can easily be removed and replaced each season. For educational purposes, these tubes can be removed and cut open to see the actual nest construction. More sophisticated bee houses show up in the form of little houses with peaked roofs or raindrop-shaped designs that hold bundled tubes made of bamboo, reeds, or rolls of corrugated cardboard.

Going the Extra Mile

People interested in observing and or increasing their local mason bee populations prefer straws or reeds that can be split open or wood trays that break apart in sections to expose the individual cocoons. In nature, many of these nests will succumb to predation or bad weather. To ensure greater success, many mason bee enthusiasts break apart the straws or reeds and harvest the cocoons. They inspect the cocoons for mites and other issues, then clean and store them in a vented shed or garage until fall. The cocoons are then refrigerated until spring so the bees' release can be timed with optimum weather conditions. When refrigerating cocoons, it's essential to maintain proper humidity that frost-free appliances don't supply. You can purchase special humidity chambers from the same companies that sell other bee housing and accessories.

Timing and Location

Prepare to put the bee house out in early spring so bees find it immediately upon hatching. If you are releasing purchased cocoons, wait for daytime temperatures to settle around 55 degrees Fahrenheit. Place them in a release box near the bee house. The box can be made of cardboard or specially purchased for that purpose. The box should have a hole the same size as their nesting tube and a small platform from which they can fly away. The hole should be above the bottom of the box and oriented so that if faces east.

Hang the block or house on an east- or southeast-facing wall or fence, between 5 and 6 feet high at about eye

This raindrop-shaped mason bee house is both decorative and functional.

level. In hot climates, place the house in filtered shade. Watch to see if bees are using it. If they're not, try moving it to another area of the garden and see if that is more attractive to them. Mason bees are done with their life cycle by the end of spring. If you harvest and clean your nests or put out new ones, you may see leafcutter bees taking up the summer shift in the same house.

Woodpeckers find bee larvae especially delectable, but other birds may be interested, too. If birds or wasps are bothering the bees, you can attach a piece of chicken wire or hardware cloth to screen it from them while leaving the bees still free to come and go.

BUMBLEBEE BARNS

Bumblebee housing can also be made at home with plans found on the Internet or purchased ready to assemble. A house consists of a small box with an opening for the cavity-nesting bees to enter. Some readymade versions have a removable see-through lid so the bumblebees can be observed in the process. Such bumblebee barns are used in greenhouses for bumblebees that pollinate tomatoes. While out in the garden with free-ranging bumblebees, this bee housing has varying rates of success.

INSECT HOTELS, BUG MANSIONS, AND MORE

They've been popular for quite a while in Europe, but insect hotels are only starting to catch on in North America. The unique and often fanciful creations offer nesting sites not only for bees but provide housing to an entire community of creatures while enriching the ecosystem in the garden.

Fashioned from found materials, recyclables, and garden trimmings, they allow for the formation of more varied nooks and crannies with different conditions both damp and dry that cater to beneficial insects and passive pollinators (the categories often overlap), such as beetles, lacewings, ladybugs, moths, and spiders, as well as a few amphibians and small mammals. Referred to as a bug stack or wildlife stack,

INSECT HOTEL INGREDIENTS

Experiment with a variety of stuffing materials to create interesting textures and patterns since these insect hotels quickly become works of art. Test the limits of your imagination to add more ingredients to this list of beautiful bug stuffing ideas.

- Small log sections
- Wooden blocks drilled with holes
- Evergreen boughs
- Straw
- Moss
- Bundles of reeds, bamboo, and branches
- Burlap folded irregularly
- Coir lining
- Pinecones
- Clay pots turned upside down or outward

- Old books
- Flat stones and tiles
- Corrugated cardboard loosely rolled
- Bricks turned backward to expose mortar holes
- Potted plants
- Shutters
- Rolls of sod
- Carpet remnants

LEFT: An eco-wall presents a multitude of nesting and sheltering sites using yard waste and other recyclables.

RIGHT: Bundles of hollow stems in this bee hotel provide places for tunnel-nesting species to raise their young.

they are often just that, made with wooden pallets or cinderblock shelving stacked on top of one another as a framework that is then stuffed with a wide assortment of what might be considered rubble and debris, but which insects find quite suitable for shelter. Not limited to straight lines when made with other materials, the frames can be circular or curving, even three-dimensional bug balls mounted on posts.

ECO-WALLS, A FRIENDLY FENCE

Insect hotels are usually standalone pieces placed about the garden, but eco-walls often work in conjunction with other hardscape elements. They work on the same principles as an insect hotel but are usually longer and sometimes shallower, stuffed with similar materials but exposed to both sides. For decorative effect, some people insert a section of stacked glass bottles to add color and transmit light in between. The materials could be changed out to reflect the seasons and holidays. Built in stackable modules, the eco-wall acts much like a hedge for the patio or between homes. Besides a great conversation starter, it's a light-hearted living wall that can divide or screen property without suggesting exclusion or hostile separation.

It should go without saying that all of these examples of manmade habitat should be placed within a convenient distance of foraging plants, water sources, and a bit of mud. It's also important that some sort of waterproof material be placed over the habitat to serve as a roof; otherwise, it may turn into a moldy mess that no one wants to either see or reside in. Lots of the natural materials used to stuff the sections of insect hotels and other habitats are purposely meant to eventually decompose. The materials are meant to "melt" as they settle and age with use and the elements. It's good for the insects as well as the design to refresh the "amenities" each year with new stocks of artful scraps, castoffs, and garden detritus. You always want to be sure that lots of pollinators and beneficial insects are excited to check in to this hotel.

BUTTERFLY HOUSES: HIT OR MISS?

That cutesy butterfly "house" shown in many designs? Researchers don't find any indication that butterflies actually fold their wings and tuck into the narrow slots; however, spiders seem drawn to the convenient crevices. There are lots of practical but decorative types of manmade habitat, but this one belongs in the category of garden décor, not cover.

Pesticides can't distinguish between this beautiful bee and a troublesome pest.

CHAPTER 8

PROTECTING POLLINATORS
WHILE CONTROLLING PESTS

HOW PESTICIDES THREATEN POLLINATORS

Go down the pesticide aisle at any garden center or home improvement store and you might be surprised at all the living things people consider pests. Look at all the cans and containers picturing devilish-looking creatures hell-bent upon making your life miserable. (With Rambo-esque brand names, these products promise to roundup, avenge, and end all pests so you can b-sure they will b-gone.) There seems to be a "cide" for just about everything, whether it's moles or mole crickets. Even weeds are included under that big umbrella label of what constitutes a pest.

For many folks, the terms "pesticide" and "insecticide" are somewhat interchangeable, but it's important to sort out what kills what. *Pesticide* is a broad term that includes many control formulas for many pests; however, insecticides kill insects, fungicides kill fungus, and miticides kill mites. You might be tempted to say that herbicides only kill weeds, according to which category (grass or broadleaf), but the truth is they don't know what's a weed and what isn't. They just kill plant growth.

And that's part of the reason pesticides can be a problem. The pesticide doesn't determine what it kills, people do. Much of the time people aren't even really sure about what it is they are trying to kill with any given pesticide. They wield a can of bug juice and spray willy-nilly hoping it rids them of the bug that is stinging them, chewing holes in their plants, or invading their homes. They spray those darn dandelions but are surprised when

Gardens with lots of biodiversity rarely need pesticides. *Cole Burrell*

According to the US Fish
and Wildlife Service,
homeowners use ten
times more chemical
pesticides per acre on
their lawns than farmers
use per acre
on their crops.

some overspray reaches their petunias, and they wither and die, too. You see, pesticides can't make moral or aesthetic judgments about what they kill; they don't distinguish between pretty flowers and pesky weeds, bad bugs and beautiful butterflies. It's funny but tragic when people insist they are only spraying for a specific pest, such as spiders, because the can says spider spray on the label, and they truly believe that's the only creature that will be targeted as they aim the nozzle.

It's hard to blame people for this naiveté since it is only relatively recently that pesticide companies are beginning to be required to list the organisms harmed instead of just the intended troublemakers. People don't budget time for a crash course in entomology when their roses are overrun with aphids or worms are munching holes in their cabbage—most look for a quick fix. While there have been warnings about harm to children and pets for some time, cautions over aquatic organisms, such as fish, and the water they live in are more recent additions to pesticide labeling. As people become more aware of the importance of pollinators, pesticide manufacturers are finally being asked to label for toxicity to bees and other pollinators so that people can make wiser decisions about when and what to spray and better yet, whether they should spray at all.

Pesticides themselves are not bad. There are certainly pests that can inflict pain, spread disease, and threaten life: fire ants, brown recluse spiders, and mosquitos are just a few that come to mind immediately. In certain situations, pesticides are great tools for saving lives and making our homes and gardens safe to inhabit. Therefore, it is important to know how they work, when their use is warranted, and the consequences.

Just as people use the blanket term "pesticide" for anything they might use to rid themselves of annoying or dangerous creatures, they also use the general term "spray" for the act of applying the pesticides. Yet pesticides come in many forms; liquid sprays are common but they also come as a granule, soil drench, seed dressing, and injection depending upon how, when, and what they need to eliminate. Some pesticides kill on contact while others need to be ingested before the desired affect of death occurs. Many that kill on contact, such as insecticidal soaps and horticultural oils, simply smother the pest or penetrate and dry out the soft outer body or exoskeleton of the organism. Some prevent maturation, eventually killing the pest. There are other pesticides that, once ingested, disrupt the digestive system or nervous system of the target, resulting in death. Some pesticides work by having the target—ants being a prime example—return and share the poison with other members of their population to kill off the entire colony.

Pesticide use wreaks collateral damage on other beneficial wildlife.

Different pesticides can be aimed at different life stages of an insect whether it's the egg, pupa, larvae, or adult, with the immature stages in some cases being more vulnerable due to not having developed characteristics such as a tough, protective exoskeleton or the ability to hop or fly away. However, once the insect is an adult, including bees and butterflies, they will have more chances to be exposed to pesticides as they move about flying and foraging over greater territory with a wider variety of possibly contaminated plant surfaces and food sources. Killing bees may be the furthest thing from a particular pesticide user's mind as they spray for some other irritating pest, but once again, pesticides don't necessarily discriminate.

PESTICIDES AND COLONY COLLAPSE DISORDER

Just as distressing as general pollinator decline is the phenomenon of colony collapse disorder (CCD), where worker bees abruptly disappear from honeybee hives. Pesticides have been pinpointed as a possible cause with one group of pesticides known as neonicotinoids being singled out as particularly worrisome in light of recent studies that try to account for CCD. Neonicotinoids (translated as "new nicotine-like insecticides") go under chemical names such as imidacloprid, dinotefuran, clothianidin, and thiamethoxam, among others. Neonics are a type of systemic insecticide, meaning that once applied, they are transported to all parts of the plant through its vascular system. No matter what part of the plant an insect may feed upon or use, whether it's the stems, leaves, petals, seeds, pollen, nectar, and even the guttation droplets, it will get a dose. And they bring it back to the hive and expose the brood as well.

Neonics are not only present in the treated plants, but can also be found in the soil they are growing in. When applied to the soil as a drench, they can persist in the environment for some time. When farmers sow treated seeds, the grinding action of the planting machinery knocks off some of the pesticide in the form of dust that can drift on the wind to other areas and contaminate other plants and soil where it settles. Untreated plants can then absorb leftover residues from that soil. These residues will also find their way into surrounding water sources.

Neonics were at first hailed as a new low-toxic alternative to more dangerous pesticides of the past. They are especially effective against sap-feeding insects, such as aphids, plus white grubs, fleas, wood-boring pests, not to mention flies and cockroaches, all creatures for which no one has too much affection. Although neonics affect the nervous systems of these insects, they were considered quite safe for bees.

Now they are linked to disturbing evidence blaming them for a plethora of problems that may explain CCD. Neonics as they are used are not defined as lethal in dosage to bees but can be accurately described as sublethal, meaning there is low-level contamination of pollen and nectar, which in the case of bees may mean merely a slower death by disorientation and subsequent death of the colony. They are now thought to be responsible for impaired foraging and feeding by disrupting mobility, ability to navigate, and olfactory learning. When affected, bees have trouble finding their way back to the hive from forage trips and communicating important information about food-gathering trips to other worker bees.

A honeybee transports a nectar/pollen mix on her legs back to the hive. Is it contaminated with pesticides?

In addition they are thought culpable for problems with reproduction and brood care by reducing food-storage capabilities and queen production, and interfering with the bees' means of temperature regulation, which is vital to the colony. They are even blamed for weakening the immune system, making bees more susceptible to the parasites and disease that were first thought to be partly answerable for CCD.

Some experts fault these studies and simply cite cold weather and poor beekeeping practices as possible explanations for the mysterious disappearances, mentioning that this is not the first time there have been such sudden and baffling losses of bee colonies in recent times. They caution about focusing on finding one silver-bullet offender when the causes of CCD may indeed be a multiplicity of factors.

It might appear that agriculture is the most egregious offender when it comes to the use of neonicotinoids. Not so fast, as it turns out, for residential landscapes and commercial greenhouses, neonics are formulated at 120 times the rate as that of agricultural usage. This potency comes as a shock to many gardeners who were sometimes even unaware of this chemical's existence much less the reliance upon it until recently. Plant breeders and growers have come to lean heavily upon the use of neonics as a kind of cure-all for a number of horticultural ills, some claiming they can't continue to produce their plants without it—at least until the pests evolve to become resistant to it, something that happens fairly quickly. Then it's on to the next one. Granted, it's much more complicated and expensive for large companies and plant breeders to wean themselves off of pesticide regimes than a home gardener with a small but precious plot to defend.

In addition, gardeners are horrified to discover that the very plants they have placed in their gardens for the benefit of pollinators may have been poisoning them instead. To avoid exposing not only honeybees but all bee species in your yard to these specific pesticides, you will need to ask lots of questions. Ask your garden centers and other plant suppliers if their nursery stock has been treated with neonics. Be careful when the garden center states that they don't use neonics since the plants could have been treated as seeds or plugs (the small, young plants nurseries often grow on to sellable size) before arriving there. If you don't get a satisfactory answer from salespeople, ask for someone in charge who can give you a definitive yes or no. Some garden centers that grow their own plants are eschewing neonics at the request of, and in many cases, the demand of their customers. However, until there comes a safe alternative to neonics, starting your own seeds (from safe sources) may be the only way for some gardeners to avoid this problem for the time being.

It becomes clearer every day that gardeners are being called to the front lines of the battle for pollinator conservation. So how do you get out of this endless spiral of chemical dependency before your garden needs rehab?

WORKING WITH NATURE

If there ever was a word in the gardening world that's worn out its welcome, it's "natural." The term natural has been overused and abused until it has little meaning. Even the idea of a garden is not truly natural. It's an artificial environment made and maintained to please people—designed for our needs over nature. It probably has a mishmash of plants from disparate, far-off places all grouped together in some kind of unnatural arrangement. The topsoil was trucked in from miles away. Chlorinated water is piped in, often from a distance, too. Some of the bugs going about their business may hale from exotic lands as well. Given this gardening version of the United Nations, you'd think it was impossible to ever hope to create a safe haven for pollinators and plants that didn't require some kind of chemical intervention.

While it will never be entirely natural, you can certainly aspire to mimic a balanced plant community, one that supports healthy relationships between plants and pollinators while being a joy to the garden's human inhabitants, too. At any given moment, there are countless complex interactions taking place between insects and other insects, insects and other animals, insects and plants, insects and fungus, mostly without our knowledge. Even the most vigilant gardener misses most of what's going on in his or her garden. Insects are busy pollinating plants, reproducing, eating plants, eating other insects and being eaten themselves, and clearing up and decomposing plant debris and other dead animals, among other vital areas of assistance. Most of the insects in the garden are engaged in some kind of positive activity, but even the ones that are misbehaving are needed to feed the ones that aren't. That's right: gardens need a certain number of bad bugs on hand all the time.

There is something of a sea change going on in the business of gardening. The attitude toward "bugs" is slowly starting to shift away from the purely negative context in which people only view them as irritating or annoying, something to be eliminated or at least avoided. They'll most likely never be considered cute and cuddly like larger animals further up the food chain. (Although have you actually looked at bees with their fuzzy little bottoms and animated antennae?) Once you have overcome the squeamish aspect—the "ew" factor you have to admit, really, because it's true—they are crucial to the garden. A garden cannot function without the essential services provided by insects and other invertebrates. Yet who thought that gardeners would eventually look for ways to make them not only safe but also comfortable and happy in their gardens?

No matter if you garden using organic or conventional methods, or perhaps a hybrid version of both, you will encounter situations where chemicals may be warranted to solve a pest problem. And for the record, chemicals can be organic or synthetic, because chemicals are everywhere, and everything is chemistry. So when the term "chemical-free" and "chemical dependency" is employed here, it's meant more to talk about chemicals with potential dangerous consequences.

If you want to reduce and hopefully eliminate the use of chemicals in your garden that have the potential to harm pollinators, the university research-based program called Integrated Pest Management or IPM is your best bet at best practices. It sounds very corporate and overly official, but simply put, IPM is an effective and environmentally sensitive approach to pest management using a combination of common sense tactics with a view to long-term control. The beauty of IPM is that it can be applied to large farms as well as home gardens even though their goals may be vastly different.

One of the basic tenets of IPM is tolerance. How much pest damage can you tolerate before your crop is destroyed or your beautiful garden is ruined? With commercial crops, the answer can be found in cold hard cash numbers. In those situations, IPM hopes to suppress pest populations below the economic injury threshold. It's not as easy to assess damage in ornamental gardens where individual opinions may fluctuate wildly as to what's ok and what's not. In edible landscapes where you count on a decent harvest, it might be simpler to calculate the cost of the tomato crop you had hoped to can for winter.

Conscientious gardeners may have to adjust their expectations and also question what really constitutes genuine harm to their carefully tended gardens. Many lifelong gardeners have been conditioned by community standards as well as advertising to react quickly to any threat to the aesthetic or hygienic wellbeing of their home landscape, to do something. Instead of reaching for the spray can at the first few holes found in a leaf or a meandering trail of ants, IPM asks you to ask yourself: is it really necessary? As mentioned in preceding chapters, sometimes the answer to creating a pollinator-friendly garden is to do little or nothing. And that's hard for many gardeners to accept. Is it really that simple?

IPM takes a proactive stance. It's important to monitor your garden. Lots of gardeners already do this, making the morning rounds with a cup of coffee to see what's blooming or about to bloom, what's ripening, and of course, how the weeds are faring. Checking for potential trouble may

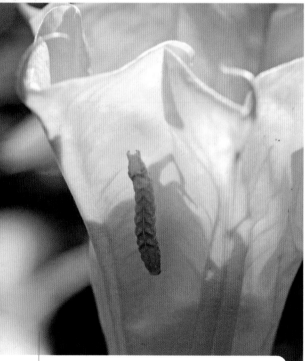

TOP: Make a positive identification of pests before deciding upon a control method, if any.

BOTTOM: Tomato hornworms are easy to identify. They eat tomatoes and other members of the nightshade family, such as this datura plant.

dampen what is normally a joyous experience, but also may put the brakes on any pest before the damage is too far gone. The positive outcome is getting to know your garden better. When you can, investigate at different times of the day or even at night. Take a magnifying glass. Look under leaves, study the bark on your trees, get down on your belly and watch what is inching and wriggling across the soil: there's a whole other secret world going on right beneath your nose. The more you know about your little ecosystem, the better prepared you'll be to take action before problems occur.

The next key is accurately identifying a pest. Just because an insect happens to be crawling on a ragged leaf at that moment doesn't mean it's the guilty party. The plant may hold several plausible suspects, so it's important to realize there may be innocent bystanders. Take pictures so you can remember all the details of the bug and the actual damage, such as the shape and location of the chewed holes, for when you do your research. Are the holes jagged or smooth? Are they along the margins or near the veins? Are other parts of the plant damaged, such as the flowers or fruit, the stems, the bark? Is there a powdery substance on the leaves, pointing instead to a fungal disease? It's also critical to ID the damaged plant since different pests often prefer different plants.

Once those clues have helped you to correctly identify the culprit, you should learn about its life cycle. This will help you to determine if the pest is a seasonal or ongoing problem, or one that will exit the vicinity soon upon completing its reproductive cycle. Often the pest is only a problem during certain parts of its life stage. If your garden only has one generation of sawfly larvae eating on your conifers, there may be time for the plants to recover that same season. And once it reaches the adult stage, that same sawfly doffs his black hat and becomes a pollinator. Sometimes if you ignore the problem or procrastinate treating it, you'll find the whole issue has resolved itself by going away!

Once you've got a positive ID, decide whether to tolerate the damage or go ahead with some sort of control. Armed with all your information, decide which method or combination of methods will be most effective as well as the best timing of the control. There are four different controls to consider.

- Biological controls use the natural enemies of the pest, such as parasites, pathogens, and predators, to control the pest and its damage.
- Cultural controls employ practices that reduce establishment, reproduction, spread, and survival of the pest.
- Physical and mechanical controls kill pests directly or make the environment intolerable.
- Chemical control is the use of pesticides.

BIOLOGICAL CONTROLS

You can purchase biological controls in many forms, such as beneficial insects, both parasitic and predatory, that you release or apply in your garden, or pathogenic fungus, such as milky spore or microscopic nematodes, to attack pests. These strategies have varying rates of success due a number of factors specific to each garden. Rather than importing these natural enemies, though, it makes more sense to encourage and welcome beneficial insects by creating conditions in your garden favorable to their continuing presence.

If you're already taking measures to make your garden pollinator-friendly, chances are you've also made it a suitable environment for beneficial insects, too. Many of their food, shelter, and overwintering and nesting requirements overlap. Planting more flowering plants will attract a greater number of beneficial insects. Planting herbs and old-fashioned, open-pollinated flowers is one of the easiest and most economical ways to lure these useful and hard-working creatures to your garden. Once they settle in, they will help maintain a healthy equilibrium of good bug/bad bug interactions, keeping pest damage to a minimal, acceptable level. Meanwhile, you'll have to keep some bad bugs around for their children to eat or they'll head off to greener— or in their case, buggier—pastures. It's hard for many gardeners to abide the fact that the answer to bug problems is more bugs, but that's the whole idea behind the self-correcting garden.

And don't forget about our feathered friends. Birds are another form of biological control. Insects make up a huge part of most birds' diets, especially when they are rearing young. So attracting birds to your garden is more than just a fascinating hobby: it's another powerful tool in your pest-fighting arsenal.

TOP: That horrible hornworm becomes a cool sphinx moth.

BOTTOM: Grasshoppers are challenging pests. Attract birds to your garden to help control their populations without pesticide use.

HEIDI HEILAND

Heidi Heiland is the chief experience officer (CEO) for her landscape gardening company, Heidi's Lifestyle Gardens, which she launched in 1979. Heidi and her team create award-winning spaces and unique experiences for residential and commercial customers and are sought out for their ability to collaborate, inform, and propose innovative opportunities to connect with the Earth, use water wisely, and practice slow-food sustainability. Heidi is a certified professional of the Minnesota Nursery & Landscape Association and its current president. She frequently presents seminars to students, colleagues, and passionate plantspeople, as well as monthly segments on a local NBC show, Kare 11's *EarthKare*.

Q. How did you become a landscape designer?

I was 17 when I started my company with my only prior experience being a water girl for a townhome development or maybe plucking dandelions from my mother's lawn. And now today you and I wouldn't pluck those dandelions, would we? I worked for my mom and her five best friends. I was an English major and took horticulture classes as well but dropped out of college since I already had my company at that point.

Q. Do many of your clients ask for garden features that support wildlife and pollinators or do you suggest them?

We have to suggest it to them at this point. They look to us to be the experts, and they are hoping to align with somebody who can take care of their landscape needs and that includes pollinator-friendly habitats or drought-tolerant approaches. We need to offer all of those solutions and have a lot in our arsenal to assist them.

Q. You and many of your clients live on lakeshores. What landscape practices do you recommend to protect water resources so important to human and pollinator health?

We believe in selling shoreline stabilization. We look at buffer zones even if it's not on a bank. We try to take advantage of existing topography to direct water back into the soil. Potentially, you may need to amend soils and create swales so that you are still managing runoff correctly. We use permeable pavers. We "body harvest" aquatic weeds, physically removing them rather than nuking the lake, so we are addressing water bodies as a whole. And certainly we are looking to not overuse fertilizers or pesticides that can end up in runoff.

Q. How do you use IPM in your landscapes?

Ensuring plant health would be first and foremost—right plant for the right place. From a permaculture standpoint, making sure the plant systems are appropriate so that the plants do most of the work themselves. We no longer do preventative spraying of insecticides or fungicides. Part of our approach today is potentially removing the host plant, like in the case of Japanese beetles; we don't really push a lot of roses.

Q. Best tips for designing pollinator-friendly gardens?

If you plant more quantity in an organized fashion in a stagger pattern, with lots of plants flowing into the next you can have a very naturalized look yet it isn't messy. There are ways you can get pollinators and natives and all of your needs met so that the neighbors don't freak out. We call it a stylized meadow.

Q. What's in your yard for pollinators?

Milkweed, *Monarda* 'Raspberry Wine', angelica, ironweed, and I'm trying not to mulch my whole yard so that the ground bees have a place to nest.

BENEFICIAL INSECTS

LADYBUGS*

There's a lot to love about ladybugs, lady beetles, ladybird beetles, or whatever you choose to call them. They are shaped like an upside-down red bowl with black polka dots. They also come in tan, orange, and yellow. Besides being cute, they are valuable predators in the garden. They feed upon aphids, mealybugs, thrips, scale, and spider mites, to mention a few. Adults will eat up to 100 aphids a day while hungrier larvae (sometimes called alligators) eat much more.

LACEWINGS*

Delicate creatures with brown or green transparent wings, their less attractive larvae are called aphid lions since they devour hundreds of aphids before reaching adulthood. In both life stages, they can be counted on to eat leafhoppers, cabbage loopers, thrips, mites, and whiteflies, among many others. When they can't find those, they will eat each other! Encourage their presence by planting flowering herbs and sunflowers.

MINUTE PIRATE BUGS*

Without a magnifying glass, they are merely black dots, but up close they are decorated with black and white triangles. They are very active predators that feast on thrips, whiteflies, aphids, spider mites, and corn earworm eggs, to name just a few. They are a bonus benefit of planting cover crops, such as alfalfa, clover, and vetch, where they like to hide.

CENTIPEDES AND MILLIPEDES

These creepy-crawly guys are naturally occurring occupants of moist, undisturbed soils. They eat slugs and snails while also acting as composters.

Adult dragonflies can eat hundreds of mosquitoes per day.

Encourage beneficial insects, such as flower flies. Their larvae are aphid-eating machines.

HOVERFLIES

Also called flower flies or syrphid flies, they are often mistaken for bees because of their black and yellow stripes. They don't linger as long on flowers as bees do, darting in and out while hovering over flowers instead. The adults are good pollinators but the larvae are great predators, gobbling up hundreds of aphids plus thrips, scale, corn borers, mealybugs, leafhoppers, and more. Encourage them with flowering herbs, mustard, Queen Anne's lace, and feverfew.

PARASITIC WASPS*

These aren't the scary kind of wasps. Instead they are tiny, non-threatening creatures that do the dirty work for you. They lay their eggs inside of caterpillars, and upon hatching, they eat the caterpillar from the inside out. Although they don't always discriminate between butterflies and bad guys, they wreak havoc on vegetable and fruit pests, such as cutworms, cabbageworms, tomato hornworms, cabbage loopers, and then some. They're attracted to parsley, coriander, and tansy in flower.

DRAGONFLIES

Like beautiful helicopters of blue, red, copper, and green or checkered black, they whiz around our gardens. They aren't residents but likely live at nearby water sources. When they visit they keep mosquitos, flies, gnats, and other flying insects in check.

PARASITIC NEMATODES*

There are bad nematodes, but these are the good ones. You can't see them, but these microscopic worms attack soil-dwelling pests such as root weevils, white grubs, slugs, snails, and maybe even Japanese beetle larvae. They are applied in a water-based solution and absorbed into the soil. They are not hardy and need to be reapplied each season in cold weather areas.

COMMON GROUND BEETLES

You may not see them but they are there lurking under logs and rocks. They are valuable additions to any vegetable garden as they chomp on Colorado potato beetles, asparagus beetles, root maggots, aphids, flea beetles, gypsy moth larvae, cabbage loopers, and a number of other pests. You can boost their presence by planting so-called "beetle banks" with hedgerows and perennials and making small woodpiles available.

   * Denotes beneficial insects that can be purchased at garden centers or through the mail. Results vary with store-bought beneficials depending upon the choice of insect and conditions in your garden. Some are better suited to use in greenhouses where the environment is more static and they are meant to target a specific pest. Success also depends upon which life stage is purchased. It may take several generations to build up a population that is effective at controlling pests. You not only want them to immediately start eating pests but also find your garden suitable for laying eggs. Some beneficials may

not survive the shipping process, so check customer refund policies before purchasing. Once you've received them, be sure to follow instructions and be ready to release or place them in the garden.

Adult ladybugs will often make like the nursery rhyme and fly away elsewhere. It's better to purchase them as larvae so they start eating pests on the spot before they can escape. Lacewing adults may also fly out of the garden so its better to buy them as eggs. It may be hard to find parasitic wasps in small enough quantities for home gardens. Garden centers rather than online sources may be a better source. Praying mantis are often found for sale at garden centers. They will devour any insect they come upon regardless of size, but since they don't congregate in groups, their effectiveness as a garden predator is questionable. Look at these fantastical stick figures as more of a fun curiosity to keep in the garden, especially for children. All beneficial insects will leave a garden and look elsewhere once the food supply is gone, so once again it's important that some bad bugs are always hanging around.

CULTURAL CONTROLS

Cultural practices involve the decisions you make about what to plant where on your property as well as what you do to maintain the garden, such as planting methods, watering, mowing, fertilizing, pruning, and weeding. You can manage those activities in ways that prevent the establishment of pests and disease, and limit pest reproduction, dispersal, and survival.

Make a good start by purchasing good quality nursery stock. Select naturally or specifically bred pest and disease-resistant varieties that won't need additional pesticide. If some of your existing plants have chronic pest and disease issues, consider replacing them. If you love bee balm and phlox for attracting bees and butterflies but don't love their annual late summer bout of powdery mildew, there are more resistant varieties available. The same goes for many other favorite plants.

Be aware of cold hardiness needs. Pay attention to soil requirements. Be careful to plant at the right depth. Match the plant to its intended location remembering the mantra of the Master Gardener, "Right plant for the right place." Given a leg up on the front end, these lucky plants have a better chance at staying ahead of pests.

Amend your soil so that plants have access to vital nutrients and minerals. Provide adequate water to ensure plant health, and avoid overwatering that may lead to rot and fungal disease. Measure and time fertilizer properly to increase growth at a desirable rate so not to create the kind of overfed, stressed plants that invite hungry pests. Suppress weed growth with appropriate spot-treatments, mulch, or hand pulling. Whenever possible, prune out diseased or damaged leaves, stems, or branches rather than turning to pesticides.

FUN FACT

Contrary to myth, earwigs don't wiggle into your ears, but they do consume large quantities of lawn pests such as chinch bugs, sod webworms, and small mole crickets.

Hand picking Japanese beetles is disgusting but effective.

WHAT ABOUT COMPANION PLANTING?

Companion planting, the idea that growing certain plants together is mutually beneficial to both, is almost as old as the hills. Tomatoes are thought to thrive when planted with basil. Beets are supposed to be more robust growing next to lettuce. The concept is rooted in fact and folklore, a combination of years of keen observation, and anecdotal evidence—but does it really work?

The answer might be that it sure doesn't hurt. Most of the gardening wisdom advocated in companion planting is based on some common sense if not scientific fact; there's usually an explanation behind the tales. With the most rudimentary companion planting, plants can be located to help each other out; a tall crop, such as corn, shades tender leafy greens, allowing you to grow them longer into the heat of summer. Versions of the well-known Native American "Three Sisters" planting arrangement sees pole beans twining around corn or sunflowers for support while rambling squash shades the soil like a living mulch.

Some crops flourish together because they share similar nutrient needs in the soil. Others, such as the beets and lettuce, do well because one is a light feeder growing above ground while the heavier feeder is seeking out sustenance deeper in the soil. Onions and their odiferous cousins, such as leeks, chives, and garlic, are counted on to repel pests. Other pungent plants, such as marigolds, geraniums, calendula, sage, and artemisia, may at least confuse them.

Finally, in old-fashioned kitchen gardens and cottage gardens, herbs, flowers, and fruit grew right there amongst the vegetables. As they bloomed they attracted lots of beneficial insects that in turn fed upon pests. In successful companion plantings, there is a synergistic effect. For example, the tomato vine is actually aided by parasitic wasps, enticed there by the blooming basil; the wasps then kill the tomato hornworm before it can gouge too many holes in tomatoes. The too-tidy vegetable garden of modern times with its repetitious weed-free rows often isolates single blocks of crops and leaves them more vulnerable to attack. A bit messier, more chaotic garden might seem untended but is probably better defended against pests. If you still crave order in the veggie plots, simply alternating rows with flowers and herbs will create the kind of biodiversity that makes the garden more self-tending and without need of pesticides.

TOP: It's said the pungent scent of marigolds repels pests. They do lure beneficial insects that indeed eat pests.

BOTTOM: Adding flowers to the vegetable garden increases biodiversity that helps control pests.

PHYSICAL CONTROLS

Physical controls, such as barriers and traps, often stop pests before they have the opportunity to cause trouble. Sticky traps can catch bothersome bugs without collateral damage to crops. However, be aware that some traps that use pheromones or hormones to lure in unsuspecting pests may actually draw more of those pests to your garden than would naturally occur. Japanese beetle traps are a prime example of a well-intentioned and seemingly sensible solution that can backfire and actually attract more of these awful pests.

You can use physical barriers, such as lightweight, floating row covers, to exclude pests from vegetable crops altogether. You can keep crops like cabbage and broccoli completely covered from transplant to harvest to eliminate damage from cabbageworms and flea beetles. However, when it comes to flowering crops, such as pumpkins and squash, be sure to only use the covers during the time when the pest is laying eggs. Otherwise, you'll want to allow bees and other pollinators full access.

There are other ways to deny access around the garden. Tender seedlings can be protected from cutworms by collars made from aluminum foil, bottomless cans, or cardboard rolls. If you have a few fruit trees in your garden, you can reduce spraying and wormholes by bagging the apples and other fruits until they ripen. It's a tedious but fulfilling task.

It may not be everyone's idea of a good old time in the garden, yet hand-picking bugs can be a worthwhile undertaking. Back in the day, you could hire kids to pick them off for a penny a pest; nowadays, you may find yourself the only one up for the job at that price. For larger pests, such as slugs, hornworms, and Japanese beetles, bring along a bucket of soapy water to dispatch your prey. If you don't like that sort of close contact, shaking off the pests is a workable alternative. Spread a sheet below the plant to catch beetles and other troublesome pests. Dispose of them so they can't return. Some people even freeze the critters and feed them to their chickens. For smaller, more numerous pests, such as the always-present aphids, a strong spray with the garden hose may knock off enough of them to keep them in check while leaving some around for the beneficial insects. Sometimes the least sophisticated methods are the most effective and satisfying.

TOP: Use row covers to protect vegetable crops from destructive larvae of the cabbage white butterfly.

BOTTOM: A backyard flock will eat bugs for chicken feed.

CHEMICAL CONTROLS

When all else fails and you desperately want to save a certain crop or individual plant, you may decide to use a chemical control. There are definitely situations when the use of pesticides is necessary or warranted. In these cases, you will want to carefully determine if the damage is merely cosmetic or could be fatal to the plant. While many trees can withstand and survive pest infestations, even through the insult of full defoliation, others can't for one reason or another. A beautiful but vulnerable mature tree that shades a home and its people may need to be treated for a particularly destructive pest. The tree might have historical significance as well. A prominent specimen shrub in your front yard, whether it's expensive to replace or maybe just a sentimental favorite, also deserves special consideration if the threat is truly great. Or you may be counting on a sizable vegetable crop to help feed your family.

After correctly identifying the pest you want to target, you should choose an appropriate pesticide, starting with the least toxic, in order to minimize harm to bees and other pollinators. Minimize harm to yourself as well; wear protective clothing, such as long sleeves, gloves, and goggles, if stated as a precaution. Make sure to follow all label directions, from accurate mixing if needed, to proper application, to safe disposal of any leftover materials. Store chemicals according to directions. Look for places in the community that take leftover hazardous materials if you don't need the entire container.

Timing of any application is imperative with regards to the life stage of the pest to prove most effective. More importantly, you want to apply the pesticide at a time of day that reduces the chance of exposure to pollinators. For many pesticides that kill on contact, it is recommended to spray only in evening, nighttime, and early morning. Honeybees will be snug in their hives and not out actively foraging during those hours. However, early morning hours could prove deadly to bumblebees and other native bees that are hardier and head out earlier in colder and damper weather.

If you or a service you are using are going to apply toxic pesticides, alert any known nearby beekeepers of your plans so they can take steps to shut their hives and protect their colonies. When using hired lawn, landscape, and tree services, ask them about which chemicals they use and what they do to minimize drift and other dangers to pollinators. If they balk at such questions or try to scare you with worst-case pest scenarios, find another service provider or get a second opinion and estimate. Lawn services may bundle numerous applications that exceed needed protection or unnecessary "prevention." Do your homework and check whether you really need all the services they may recommend; see if you can buy their services a la carte to reduce the number of applications. Not only will you save money but vital pollinators as well.

Remember that sometimes after trees, lawns, or garden plants have been treated for a particular pest, its natural predators are killed off if the process is too disrupting to the equilibrium of the landscape. This can result in a boomerang effect with a bigger surge of the same pests the following season and the start of a vicious cycle of repeat applications.

HERBICIDES

When people use herbicides, they often don't make the connection between plant and pollinator. After all, they are spraying weeds not insects, right? Not counting the incidental insects that may be among the weeds, any pollinators in the area are negatively affected since many so-called weeds are food sources for bees and host plants for butterflies. The damage may be indirect but is just as harmful. As farm acreage has increased, roadside forage that accounts for a huge part of a pollinator's diet has disappeared. Plain and simple, herbicides reduce food availability by reducing the number of flowering plants.

JUST BECAUSE IT'S ORGANIC...

There are lots of organic alternatives to synthetic pesticides. But the organic designation doesn't necessarily mean it's safe for people, pets, pollinators, or the environment. Only a few organic solutions are without some level of toxicity. Some experts even question the use of garlic and similar pest repellents because they may mask floral aromas, confusing and discouraging bees from visiting treated flowers. Other insecticides are innocuous for certain life stages while deadly to others either for adult forms or larvae. In other cases, careful timing of application is the key to reducing pollinator harm.

Bacillus thuringiensis: Known by its nickname "Bt," bacillus thuringiensis is a naturally occurring bacterium that is applied as a dust or liquid. It acts as a stomach poison, rupturing the pest's guts. It doesn't affect organisms that are not actively eating the treated leaves, so it is considered safe for bees. It's effective against cabbageworms and tomato hornworms (the latter of which do turn into fascinating sphinx moths). Use with care around host plant-herbs, such as parsley, dill, and fennel, where desirable butterfly caterpillars may be dining.

Beauveria bassiana: This naturally occurring pathogenic fungus is used as a mycoinsecticide against the holy trinity of garden pests; thrips, aphids, and whiteflies. It infects and penetrates the cuticle of the pest, causing death, and it is sometimes combined with other insecticides. It is highly toxic to leafcutter bees and also harmful to bumblebees.

Boric acid: This is a common ingredient of ant and cockroach controls. It has the potential to kill bees on contact, but since it is usually used only in and around structures, it's unlikely bees will encounter it.

Diatomaceous earth: DE is touted as an organic for a number of home and garden pests from bedbugs to flea beetles. It's a very popular recommendation for slug and snail control. Made up of tiny diatoms that form silica, the microscopic sharp edges kill many soft-bodied pests upon contact while it also readily absorbs into the exoskeleton of others, causing dehydration and death. Although evening and late night applications may avoid exposure to foraging bees, the residual dust may later become trapped in the hairs of bees and brought back to the colony where it can harm both adults and larvae. Humans should be careful to avoid breathing its dust.

Horticultural oils: Horticultural oils are used to smother many bothersome pests. They are a widely used pest control for fruit and nut orchards where foraging bees may be working. They are deadly to bees on contact but safe when applied in evenings, nights, and early morning or when trees are dormant and bees are not active.

Pyrethrins: Derived from chrysanthemums, they sound safe but are actually highly toxic to bees and persist in the environment for up to seven days.

Copper, Copper sulfate, Lime sulfur: Used as fungicides in different combinations, they have been shown to have various negative impacts on bees and their reproduction. Use with care where bees may be present and also in nesting areas.

Neem: Neem is a botanical extract that disrupts the hormonal systems of chewing and sucking pests. It isn't an immediately noticeable solution since its effect is to prevent maturation of the insect. It has no ill effects on adult bees but should be only applied when bees are not present. There is the potential for them to carry it back to the colony where it would endanger larvae.

Learn to appreciate and tolerate dandelions. The bees will thank you.

Additionally, genetically modified crops can now withstand applications of herbicides that decimate vulnerable wildflowers and weeds that are the mainstay of pollinators. Indeed, modern agricultural practices bear the lion's share of the blame for this, but the home gardener isn't off the hook by any means. And gardeners may be the only hope for restoring millions of acres of lost habitat, one backyard at a time.

After determining which weeds you find bothersome, look at them from a bee or butterfly's perspective. Sometimes those weeds, such as dandelions, are the only food source for miles, especially during the first weeks of spring. Are they really that terrible looking or is it years of conditioning that makes you want to murder them the minute they emerge? Use the same IPM criteria as for pesky critters: decide how many and which weeds you can tolerate or grow to appreciate. Maybe you can handle more in the backyard while keeping the front yard neater. That's a start. For home gardens with manageable numbers of weeds, hand pulling is still the least harmful method. It's good exercise, too. Some people even tell of enjoying the process, reporting a sense of satisfaction, while others find it a meditative act. If that sounds doubtful to you, consider only spot treating especially tough weeds instead of using the heavier-handed shotgun approach.

Sometimes herbicides are the only cost- and time-effective way for a homeowner to remove noxious or obnoxious weeds. Glysophate herbicides are recommended as the best way to remove large swaths of weeds that have taken over or before replacing an area with a new landscape or garden. There are studies both pro and con debating possible harmful effects and persistence in the area treated, along with runoff issues.

To minimize harm to other plants in your yard, the neighbor's yard, and beyond, always spray during optimal weather days. Avoid dewy days when pesticide residues remain longer on wet foliage. Wait out any windy days for calmer conditions. Herbicide drift is often hard to see until the withering damage is done. Always calibrate equipment and adjust nozzles before setting out to spray. Spot treating is more accurate with sprays that use a more visible, foamier consistency. Whenever possible try to wait for that window between when the weeds are done flowering but before they are actively dispersing seeds.

There are alternative strategies for large areas of weeds that involve more muscle and determination. Repeated tilling can sometimes take out enough vegetation so that subsequent weed growth can be managed with manual pulling or only spot treatments. Solar sterilization is accomplished by covering the area with black plastic that kills weeds and seeds to a shallow depth. It takes several months to a season of warm weather to complete the process. No matter which weed-killing method is used, hopefully the replacement plants in the new landscape contain as much or more food value for people and pollinators once they are established.

Be an advocate for bees and other pollinators. There are lots of ways to help.

CHAPTER 9

ADVOCATING FOR POLLINATORS

EDUCATE AND ADVOCATE

By the time you reach this chapter, you've already learned a lot about encouraging and supporting pollinators in your garden. But don't stop in the yard. Keep on learning! Continue to observe and enjoy pollinators everywhere you go! Visit botanical gardens and arboretums. Pay attention to nature's own design tips when you explore wild areas. Join a conservation group.

Then think seriously about sharing all this powerful knowledge with your friends, neighbors, and others. That's the premise and hope of this book. One pollinator-friendly garden begets another one until there are hundreds and then conceivably thousands of them forming valuable wildlife corridors that link habitats together, making food, shelter, and nesting sites available for countless species of pollinators.

There's no better way to dispense this information than one-to-one while you're out there tending your garden. Once you have a pollinator-friendly garden, people will take notice. They'll see that your garden has more flowers, more birdsong, more life. At first they may not ask questions, but you'll see them slowing down as they drive by, pausing as they walk by. They'll smile. You'll smile back. After all, it should be just as welcoming to people as it is to pollinators.

Here's the chance to play the role of sidewalk evangelist for pollinators. But rather than pontificate about your gardening principles, point out a bumblebee; take them closer to watch it for a minute. Identify a bloom or two. Ease them into it.

Bee watching is the new bird watching.

FUN FACT

Honeybees living in the White House kitchen garden produce honey for state dinners and diplomatic gifts as well for use in crafting specially made beer.

Mention how the garden not only attracts these fascinating creatures but also sustains them through their entire life cycle, providing your garden with beauty as well as their valuable pollinator services.

Of course, you can't be out there all the time, although you may want to be. So consider certifying your yard as an official habitat. Depending upon your garden's focus, you could certify it as a Backyard Habitat with the National Wildlife Federation, as a Monarch Waystation with Monarch Watch, or another of many such programs. Pollinator habitat and bee-safe certification programs are popping up all over the country. These programs often provide plant lists and other tips to help you fulfill their requirements. Once you're certified, you can donate a small amount to obtain a small yard sign that explains the criteria for establishing your habitat.

The sign is great for those passersby who are too shy to ask questions. It will pique the curiosity of those driving by. It's fun to watch them stop, look around, and then read the sign and then look up and around again at your yard. You can actually see the wheels turning as they regard it.

Social media is another great way to share information about pollinator conservation. You can inspire and inform others with pictures of your pollinator-friendly garden, news about pollinator-focused events, and more. Be careful, though, not to post too much, too often, since you don't want people to mistake your enthusiasm for spam! Social media is also a place where misinformation on pollinator issues can spread fast without the benefit of editorial oversight or basic fact checking. Before you hit the "share" button or even the "like" button, consider the source. Does the story come from a reputable organization or individual? Is it a website seeking to sell something? Beware of the numerous online petitions that proliferate on Facebook but sometimes only seek your personal information for marketing rather than true environmental change.

Pollinators are in the news as well. It seems not a week goes by that you don't see something about bees or butterflies anymore. As this book was being written, the US Fish and Wildlife Service announced it will spend $2 million to grow milkweed and other butterfly-friendly plants along the monarchs' main migration route from Minnesota to Mexico. Partnering with the National Wildlife Federation and the National Fish and Wildlife Foundation, their aim is to restore 200,000 acres of habitat in the spring breeding grounds of Texas and Oklahoma to the summer breeding grounds in the Corn Belt. The lofty goal is to eventually bring back the monarch population to one billion after fewer than 50 million made it to Mexico

TOP: Welcome potential pollinator-friendly gardeners with gentle persuasion.

BOTTOM: Your garden will inspire others to create their own wildlife habitat.

in 2013. Finally, pollinator summits of all kinds and sizes are being called around the nation as more people become more aware of the issue.

Whether you get involved on a local, regional, or national level, you can make your voice heard in support of pollinators. Whether writing a letter to the editor, speaking out at your local park board, or addressing your legislators, be clear and concise about your concerns. Well-intentioned but rambling odes and speeches test peoples' attention span and turn off potential supporters to your cause. Limit your concerns to a few key points and state them with confidence. Show your passion but maintain your credibility by using research-based information and a sense of optimism.

JOIN UP

Creating awareness of the critical role that pollinators play in the health of people and our planet is the first step toward real action to conserve them. Among the many worthy organizations that seek to promote the cause, consider joining:

- Xerces Society
  www.xerces.org
- Monarch Watch
  www.monarchwatch.org
- Ladybird Johnson Wildflower Center
  www.wildflower.org
- Pollinator Partnership
  www.pollinator.org

CITIZEN SCIENCE PROJECTS

These valuable projects are only possible through the efforts of concerned and curious "citizen scientists" who make observations and gather data in the field for the purpose of broadening studies of pollinators and the plants they rely upon. You can volunteer for citizen science projects that ask for only 15 minutes of your time—or much more, if you want, depending upon your schedule and level of interest.

They may involve simply counting species in your backyard for a particular time period, or there may be more involved activities such as tagging species or observing and recording animal behavior, plants, or weather. Find out about projects in your area through universities, conservation societies, extension agencies, local nature clubs, and continuing education programs.

A pollinator-friendly garden will pique the curiosity of passersby.

RAISING MONARCHS

Along with the fresh food-growing revolution, gardeners have been on the leading edge of a new livestock movement with chickens, ducks, bees, goats, and other four-legged critters appearing in the garden plot. There's another type of farming that you might say is "creeping" up in popularity, rearing monarch butterflies from egg to wing. Whether armed with pickle jars or professional kits, there are a surprising number of conservation-minded folks raising monarchs at home, helping the cause one butterfly at a time.

It's a simple and rewarding activity, whether you join the ranks as a curious conservationist, citizen scientist, or classroom educator. It's a chance to observe firsthand the entire circle of life within the metamorphic process. With a minor investment of spare change and time, you can witness one of the most incredible natural phenomena up close and personal while helping out a species that's not yet on the endangered list, but whose migratory lifestyle is definitely threatened.

The number one "pro-tip" for the project is to avoid disease by starting with sterilized containers and maintaining cleanliness throughout the process. Use a 10-percent bleach solution for any container you plan to use, no matter if it's the plastic clamshell from yesterday's lunch or a store-bought setup.

But first go out and see if you can find some eggs. It's not as easy as it seems. Look for adult monarchs visiting milkweed plants, watching in particular for females pausing to lay eggs on or underneath leaves. This can be a challenging quest since they usually only lay one little pearl-like egg per plant to make sure the larva has enough to eat. Don't give up too soon. Some "monarch ranchers" say the top third of the plant is more likely to yield an egg. At this point, however, you might want to know there are sources for mail-order monarch supplies with eggs, caterpillars, and even chrysalises available for purchase. (See Resources on page 170.)

When you find an egg, collect that leaf or entire stem and place it in water back home. You'll want to keep this milkweed and subsequent supplies of milkweed fresh. Some people simply put the milkweed stem in a vase of water and allow the larva to hatch and crawl on the plant out in the open during the first and second instar stages when it's rare for the caterpillar to stray from its food source.

The size of the container is dependent upon how many caterpillars you intend to rear. A deli or sandwich container is fine for one as long

TOP: The monarch chrysalis looks like a delicate piece of jewelry.

BOTTOM: The monarch emerges from its chrysalis.

as it is large enough for the emerging adult butterfly to stretch its wings and move about before it's released. If you need to transfer a tiny caterpillar use a soft paintbrush to move it to the larger container. In their later stages, you can handle them gently if needed.

You'll want to go bigger if you have aspirations of raising multiples. Many people use plastic buckets of various sizes with mesh netting as a "roof" secured with rubber bands or ties. Ice cream buckets and wedding tulle fabric are a favorite combination. Seasoned experts advise against metal lids with holes that can encourage condensation and mold and don't provide enough air. It's important to be able to easily reach into the container regularly while refreshing the food supply or cleaning out the frass, which is a fancy name for caterpillar poop. You may have an old aquarium around; it's a perfect habitat that allows you to observe the metamorphosis from all angles. A screen over the top provides a place for the caterpillar to pupate and hang in chrysalis form.

The market has responded to the growing popularity of this monarch-rearing activity with purpose-made mesh containers ranging from around a foot-square cube to taller setups going from 2 to 3 feet tall up to 6 feet tall. These larger types are designed so you can place a potted plant—milkweed, in the case of monarchs—in the protected container as a fresh food source. The fine gauge mesh allows for adequate air circulation while it prevents ants and wasps from entering and tiny larva from escaping. They feature a zippered entry on one side and a vinyl observation window on the other, and they collapse flat for easy storage when not in use. There are also mesh socks or sleeves for sale that can be used to protect caterpillars found in place outside. (See Resources on page 170 for more information.)

After your egg hatches and the miniscule instar starts munching on the milkweed, it's imperative to keep more milkweed on hand, either storing it in the fridge or picking fresh every few days. Wrap the stems in moistened paper towels to help this along. If you're raising multiples, you'll be surprised at how much they go through. And what goes in comes out: clear frass regularly. Paper towels or newspaper on the bottom of the container makes this job easier. In spite of your best intentions, some caterpillars may become diseased. If you notice them turn black and elongate, remove them promptly from the container. If possible you may want to temporarily relocate your remaining livestock and disinfect the container with a 10-percent bleach solution for five minutes. Rinse well and dry before moving remaining healthy caterpillars back in.

Once the caterpillars molt through to the final instar, they will be ready to pupate. They'll crawl to the top of the container, attach themselves with a silken thread, assume a J-shaped pose, and shed their skin for the last time. The resulting chrysalis is nothing short of amazing: a beautiful, jade-colored case with shiny gold edging and dots that resembles an elegant earring. After 10 to 14 days, the chrysalis will turn transparent and the orange and black wings will appear to darken the vessel. The whole event happens fast, usually around mid-morning. The emerging butterfly will need to inflate its wings, pumping blood into the veins. If the weather is warm enough, the butterfly can be released at the end of the day. If you need to keep the butterfly inside longer for bad weather or observation, feed with watermelon or honeydew. You can provide a 20-percent honey-water solution or Juicy Juice-brand fruit juice. Fresh-cut flowers can supplement as well.

When it comes time to release the butterfly, you'll probably feel a range of emotions from pride to awe knowing you've assisted such a miracle of nature.

PROJECTS FOR BEES

- The Great Sunflower Project (www.greatsunflower.org) asks citizens to record pollinator data from their gardens, schools, and parks. They are especially interested in pollinator visits to sunflowers. They have the largest single body of information on bee pollinator service in the US.
- Bumble Bee Watch (www.xerces.org/bumble-bee-watch/) seeks volunteers to track the species of North American bumblebees they encounter to better follow their status and inform more effective conservation efforts.
- Vermont Bumblebee Survey (m.vtecostudies.org/vtbees/index.html) aims to document abundance and distribution of bumblebees and eastern carpenter bees across Vermont. The data will be given to the public and policymakers for the purpose of making conservation and management decisions.
- Bumblebee Brigade (www.wyomingbiodiversity.org/public-programs/summer-pollinators/bumble-bee/) is a group of people who record bee sightings to better learn about the different species living in the state of Wyoming.
- Bumble Boosters at University of Nebraska-Lincoln (bumbleboosters.unl.edu) uses innovative, research-based teaching methods and cutting edge technologies to promote pollinator conservation.
- Bee Spotters (beespotter.mste.illinois.edu) collect data on bee sightings in Illinois and Missouri to help contribute findings for a nationwide effort of baseline information about bee populations.

PROJECTS FOR BUTTERFLIES

- Monarch Watch (www.monarchwatch.org) has many ways to get involved through tagging and migration monitoring, monarch rearing, monarch waystations, and milkweed conservation efforts among others.
- Butterfly Census (www.naba.org) is sponsored by the North American Butterfly Association and asks volunteers for one-day butterfly counts in spring, summer and fall.
- Monarch Larva Monitoring Project (www.mlmp.org) is connected with the University of Minnesota. Its volunteers collect data to advance better understanding of how and why monarch populations vary in time and space. It seeks to aid in conserving monarchs and their threatened migratory phenomenon.
- Monarch tagging goes on in late summer at the Great Smoky Mountains Institute at Tremont in Tennessee. The group tags monarchs to learn more about their migration and population status. Find more information at www.gsmit.org/CSMonarchTagging.html
- The Vanessa Migration Project (www.birds.cornell.edu/citscitoolkit/projects/vanessa/) is a Cornell University program that has citizen observers report their sightings of four butterflies of the genus Vanessa in order to monitor their seasonal distribution, track migrations and outbreaks, as well as study effects of weather and climactic patterns on butterfly activity.
- "Monarchs in the Classroom" (monarchlab.org) provides a wide variety of materials and professional development opportunities for teachers, naturalists, and citizens throughout the US. The program seeks to help students learn more about the natural world through the life cycle and migration of the monarch butterfly.

PROJECTS FOR HUMMINGBIRDS

- Hummingbirds at Home (www.hummingbirdsathome.org), sponsored by the Audubon Society, asks you to join and learn more about hummingbirds and how to protect them.

Become a bee spotter for citizen science projects.

DR. REBECCA MASTERMAN

Rebecca Masterman first worked for the University of Minnesota Bee Lab as an undergraduate in 1992, and returned in 2012 as the Bee Squad's associate program director and coordinator. Becky graduated from UMN Twin Cities with a BA and then obtained a Ph.D. in Entomology. She is currently studying the neuroethology of honeybee hygienic behavior under the direction of Dr. Marla Spivak.

Q. How did you come to study bees?

My major was history, and I decided I didn't know enough about biology. I looked at getting lab experience and ended up in the entomology department. I became fascinated by insects. I first worked in a lab rearing corn borers, and when that job ended, Dr. Spivak hired me and introduced me to bees. I got really lucky: I was in the right place at the right time. I was hooked very, very quickly.

Q. The Bee Squad sounds like a band of superheroes fighting for pollinators! What exactly does the Bee Squad do?

The Bee Squad was developed for two reasons: to help beekeepers and to help people and organizations help bees. We're really here to teach them the basics, how to monitor colonies, how to manage colonies. We're also trying to be a source of information, as far as how people can support bees in their yards with what they are planting. And we have the Hive to Bottle program where a Bee Squad beekeeper takes care of bees on your property while they are busy taking care of you and the environment by pollinating food and native plants.

Q. So you make house calls?

We do. Besides Hive to Bottle, we also have a Home Apiary Help program that helps with emergencies if somebody has a situation they don't quite know how to handle.

Q. Tell us more about Hive to Bottle.

The Hive to Bottle program is a surprise to us, as far as how successful it is and how much good it is actually doing for the bees. We are seeing tremendous engagement, not just with the families and friends of the families who have bees, but from the companies and organizations that have bees and use them to talk about bee health issues with their employees as well as their customers.

At first they were worried about the logistics, the extra work. But they absolutely find that the employees care so much about what's going on with that specific colony they change things not only at the business, they change practices. Some tell me that they went home and planned a bee garden because they now realize that bees need more food. I've been told by somebody they were going to spray for weeds, then they saw water next to the place where the weeds were and realized then the bees might be drinking herbicides. So they made a change as far as chemicals used for landscaping. If that's not an amazing story of change, I don't know what is. It's exciting because that is how things get done, when people become personally invested in the problems.

Q. Are there programs like this in other parts of the country?

We're in talks with a couple of other locations to expand the Bee Squad. We'd like to do it though different universities and in places where it makes sense and there is a demand for it.

Q. What's in your yard?

I actually have seven bee-friendly yards because we have rental properties. We have an extensive amount of clover in all of them. We have fruits planted, raspberries and cherries. We grow lots of sedum for fall feeding, lots of sunflowers, too. And there's a garden in every yard so people can plant their own vegetables.

PROJECTS FOR BATS

- Bat Detective (www.batdetective.org) asks citizens to help record bat calls in order to learn more about the status of their populations. Traditional visual surveys are hard to do because bats are small, nocturnal, and hard to catch, but much can be learned through these acoustical methods.

RELATED TO PLANTS AND POLLINATORS

- Project Budburst (budburst.org) gathers data on the timing of leafing, flowering, and fruiting of plants in order to learn more about the responsiveness of individual plant species to a changing climate.
- National Phenology Network (www.usanpn.org) collects standardized ground observations of teachers, students, and volunteers to promote a broader understanding of plant and animal phenology and its relationship with environmental change.
- Tracking the Wild Invasives (www.birds.cornell.edu/citscitoolkit/projects/trackinginvasives/) is a USDA-supported project that aims to better understand the spread of invasive plants in forested parklands of southern New York and northern New Jersey that have high conservation value and high levels of public use.

GARDEN SEEDS

These seed companies have long promoted pollinator-friendly gardening practices with seeds that produce plants beneficial to bees, butterflies, and hummingbirds. They are noted with special icons in many of their catalogs and websites. In some cases, they sell special seed mixes for pollinator gardens.

Renee's Garden Seeds
www.reneesgarden.com

Select Seeds
www.selectseeds.com

Johnny's Selected Seed
www.johnnyseeds.com

John Scheepers
www.johnscheepers.com

Seed Savers Exchange Demonstration Garden
www.seedsavers.org

The Cook's Garden
www.cooksgarden.com

Pinetree Garden Seeds
www.superseeds.com

NATIVE PLANTS/SEEDS

These companies have been on the forefront of native plant gardening and supporting pollinators. In addition to offering excellent customer service, their websites present a wealth of information for the pollinator-friendly gardener.

Wildflower Farm – Wildseed Farm
www.wildseedfarms.com/home.php
Store: Fredericksburg, Texas; and online

Prairie Moon Nursery
www.prairiemoon.com
Nursery: Winona, Minnesota; mail order and online

High Country Gardens
www.highcountrygardens.com
New Mexico, online only

Nashville Natives (Plants for Pollinators)
www.plantsforpollinators.com
Nursery: Fairview, Tennessee; and online

Plant Delights Nursery
www.plantdelights.com
Nursery: Raleigh, North Carolina; mail order and online

Mid Atlantic Native Plants
www.midatlanticnatives.com
Nursery: New Freedom, Pennsylvania; and online

Wildflower Farm
www.wildflowerfarm.com
Coldwater, Ontario, Canada; phone, mail order, and online

FRUIT AND NUT TREES

A vast selection of fruit and nut trees plus hard-to-find native varieties can be sourced from these companies.

Oikos Tree Crops
www.oikostreecrops.com
Kalamazoo, Michigan, online only

Stark Bros.
www.starkbros.com
Nursery: Louisiana, Missouri; mail order and online

BEES AND BEEKEEPING SUPPLIES

Just about anything you could ever need for beekeeping.

Mann Lake
www.mannlakeltd.com
Hackensack, Minnesota; Woodland, California; and Wilkes-Barre, Pennsylvania; and online

NATIVE BEES AND NESTING SUPPLIES

These companies champion the cause of native bees. Their catalogs and websites offer great advice and tutorials on mason bees, leafcutter bees, and bumblebees.

Crown Bees
www.crownbees.com
Washington state; online only

Territorial Seed Company
www.territorialseed.com
Oregon state; online only

Raintree Nursery
www.raintreenursery.com
Garden center: Morton, Washington; and online

Peaceful Valley Farm Supply
www.groworganic.com
Store: Grass Valley, California; and online

RECOMMENDED READING

Attracting Native Pollinators, Protecting North America's Bees and Butterflies, The Xerces Society Guide

Bringing Nature Home, How You Can Sustain Wildlife with Native Plants, by Doug Tallamy

Garden Home, by P. Allen Smith

Great Garden Companions, A Companion Planting System for a Beautiful, Chemical-Free Vegetable Garden, by Sally Jean Cunningham

Hummingbirds and Butterflies, Bird Watcher's Digest, by Bill Thompson III and Connie Toops

Insects and Gardens, In Pursuit of a Garden Ecology, by Eric Grissell

Landscaping with Native Plants of Minnesota, by Lynn M. Steiner

The Living Landscape, Designing for Beauty and Biodiversity in the Home Garden, by Rick Darke and Doug Tallamy

Milkweed, Monarchs and More, by Karen Oberhauser and Michael A. Quinn

Native Alternatives to Invasive Plants, C. Colston Burrell

Pollinators of Native Plants, by Heather Holm

Second Nature, A Gardener's Education, by Michael Pollan

Taming Wildflowers, by Miriam Goldberger

USEFUL WEBSITES

Illinois Wildflowers
www.illinoiswildflowers.info
This site goes beyond Illinois with helpful information on wildflowers and native plants common to larger areas of the US. Search for prairie, wetland, woodland, and savanna plants, plus their relation to flower-visiting insects and plant-feeding insects.

Minnesota Wildflowers
www.minnesotawildflowers.info
This website provides helpful information for identifying wildflowers found in Minnesota and other northern states searchable by name, color, and bloom time.

Missouri Plant Finder
www.missouribotanicalgarden.org/plantfinder/plantfindersearch.aspx
Missouri Botanical Garden's searchable database of 6,800 plants, both native and introduced plants common to the US. Search for origin, growing conditions, landscape uses, cultivars, pollinator relationship, and more.

Native Plant Database
www.wildflower.org/plants/
Ladybird Johnson Wildflower Center's database contains over 8,000 plants native to North America. Search for plant characteristics, bloom information, distribution, growing conditions, wildlife benefits, supplier directory, and more.

PLANTS Database
plants.usda.gov/
Natural Resources Conservation Service's has regionally specific standardized information on native plants, invasive plants, endangered species, and pollinators, along with range maps and other references.

University of Minnesota Bee Lab
beelab.umn.edu
The Bee Lab's site contains a wealth of information on both honeybees and wild bees, emphasizing practical solutions and the latest research on environmental issues facing them.

University of Minnesota Monarch Lab
monarchlab.org
The Monarch Lab's site is a resource for those interested in monarch conservation, biology and research, education and gardening. Information is available on the Monarch Larva Monitoring Project.

ACKNOWLEDGMENTS

I've been gardening and writing for as long as I can remember, but it was only about 15 years ago that I realized I could combine the two. I am proof that it's never too late to write (and photograph) your first book.

I'd like to thank Voyageur editor Thom O'Hearn for his patience and guidance helping me through the learning process, and art director Cindy Laun for her help as well.

I am grateful to P. Allen Smith for writing the foreword to this book. His knowledge, inspiration, and sense of stewardship made him my only choice. I want to thank him for inviting me to Moss Mountain for those great Garden2Blog events, which helped to bolster both my confidence and career as a garden writer.

I want to thank Cole Burrell, George Coombs, Heidi Heiland, Dr. Rebecca Masterman, Dr. Donald Mitchell, Dr. Karen Oberhauser, Dr. Marla Spivak, Dr. Chip Taylor, and Benjamin Vogt for giving generously of their time and expertise.

Thanks to Elaine Evans, Yuuki Metreaud, and Jonathan Neal for their info and advice on many bee questions.

Thanks to Peter Morrow for allowing me to photograph his beehives and the buzzing inhabitants. Thank also for the delicious honey!

Thanks to Meleah Maynard (monarch rearing), Bob Wolk (eco-wall), Barb Gasterland (urban prairie), Janet and Janice Robidoux (native gardens), and Donna Hamilton (urban flower gardens).

Thanks to Kathy Johnson, Nancy Leasman, and many other Minnesota Master Gardeners for their insights into rearing monarchs.

Thanks to Gallery Espresso in Savannah, Georgia, for the kindly baristas, endless cups of tea, cozy couches, and reliable wi-fi that saw me through the writing process while escaping the long winter in Minnesota.

Thanks to my husband Tom for being proud of me and for not complaining about the "book" scattered over the dining table for most of the year. Thanks to my children, Hannah and Will, for their love and support.

Finally, thanks to my mother for always making sure I noticed the wildflowers on the side of the road. And to my father for making sure I saw the birds in the sky.

ABOUT THE AUTHOR

Rhonda Fleming Hayes is a lifelong gardener and award-winning garden writer whose experience spans from her native California and all around the south, Midwest, and even to England. She pens a garden column in the *Minneapolis Star-Tribune* and *Northern Gardener* magazine. She has won the Garden Writers Association silver award and is widely published. She is a Master Gardener, public speaker, and always encourages her audience to plant for pollinators.